Exploring Science through Young Adult Literature

Exploring Science through Young Adult Literature

Edited by
Paula Greathouse,
Melanie Hundley, and
Stephanie Wendt

ROWMAN & LITTLEFIELD
Lanham • Boulder • New York • London

Published by Rowman & Littlefield
An imprint of The Rowman & Littlefield Publishing Group, Inc.
4501 Forbes Boulevard, Suite 200, Lanham, Maryland 20706
www.rowman.com

86-90 Paul Street, London EC2A 4NE, United Kingdom

Copyright © 2023 by Paula Greathouse, Melanie Hundley, and Stephanie Wendt

All rights reserved. No part of this book may be reproduced in any form or by any electronic or mechanical means, including information storage and retrieval systems, without written permission from the publisher, except by a reviewer who may quote passages in a review.

British Library Cataloguing in Publication Information Available

Library of Congress Cataloging-in-Publication Data Available

ISBN 9781475866360 (cloth : alk. paper) | ISBN 9781475866377 (pbk.) | ISBN 9781475866384 (epub)

Contents

Introduction   vii
Paula Greathouse, Melanie Hundley, and Stephanie Wendt

1. Thirsty for Science: Exploring Water Systems, Water Conservation, and Drought through *Dry*   1
   Michael DiCicco and Chris Cook

2. Climate Change is *A Hot Mess*: The Human Impact on Earth Systems   23
   Shelly Shaffer and Kathryn Baldwin

3. Countering "Plant Apathy": Using Kenneth Oppel's *Bloom* as a Motivating Tool for Teaching Plant Science to Students   45
   Katharine Covino and Erin Rehrig

4. Making Botany Magical: Teaching about Plants with *This Poison Heart*   61
   Julie C. Baker, Shawn E. Krosnick, and Kelly Moore

5. Exploring Nature and the Nature of Scientific Inquiry: Reading *The Evolution of Calpurnia Tate*   77
   Amy Palmeri, Emily Pendergrass, and Heather Johnson

6. Past and Future Plagues as Windows into the Present: Reading *A Death-Struck Year* to Teach about Diseases and Immunity   95
   David Nurenberg and Ben Lawhorn

7. Reading *Ringside, 1925*: Text Support for Teaching Evolution   115
   Frances A. Hamilton and Dana L. Skelley

8	Studying Genetics and Ethics through Young Adult Literature: How *The Gardener* Can Harvest Student Engagement in Biology *Janine J. Darragh, Ashley S. Boyd, and Kristina L. Podelnyk*	133
9	Hungry for More: Exploring, Experimenting, and Engineering with *The Hunger Games* *Leslie Suters and Kristen Pennycuff Trent*	157

Index	175
About the Contributors	177
About the Editors	183

Introduction

Paula Greathouse, Melanie Hundley, and Stephanie Wendt

In our standards-driven courses, making concepts relevant and meaningful to students while meeting the requirements of the curriculum can be a challenging task. Educators often rely on assigned textbooks and resources publishers create to determine what needs to be learned and how (Budiansky, 2001; Daniels & Zemelman, 2004). But what is presented in these mediums is often a single perspective of concepts and a lack of connection of the concepts to the lives of the students who are expected to learn from them. An unintended consequence of this is the wavering of engagement and motivation to learn, as students see little to no relevance of the content to their lives (Moore et al., 2000). In an effort to move students toward a deeper understanding of content area concepts, it is important that teachers offer multiple texts and opportunities for students to "see" these topics in their world. One way to accomplish this is through the inclusion of young adult literature (YAL) in the content area classroom.

YAL features characters, plots, and themes that are relevant to the lives of secondary students. Additionally, many young adult novels also spotlight content area concepts. This connection holds the potential for YAL to not only influence motivation and engagement, but when utilized as a complement to the curriculum, YAL can extend students' understanding of complex concepts by offering them opportunities to imagine them outside of the classroom. In other words, by incorporating YAL in a content area classroom, students can imagine complex concepts beyond the textbook.

When including literature that spotlights specific content topics, it has been noted that students tend to show improvement in reading comprehension, vocabulary development, and enthusiasm for reading (Wallace & Coffey, 2016). By helping students develop these skills, they hold the potential for becoming more comfortable discussing their understanding of content. As such, positioning students as readers in this way helps develop critical literacy, as concepts are explored and literacy is practiced at deeper levels.

One of the goals of science education is to promote scientific literacy, defined as the "ability to engage with science-related issues, and with the ideas of science, as a reflective citizen" (Programme for International Student Assessment, 2015, p. 5). To achieve this, students need to master the act of following scientific principles and balancing this application to real-life situations. Selecting the right material can be a challenge in itself, but it is possible if we expose students to a variety of texts (Wellington & Osborne, 2001). The aim of this book is to offer science teachers instructional approaches in including YAL as a supplemental text in their science classroom as a way to provide students opportunities to develop scientific knowledge and literacy skills while also "seeing" science in their world. In other words, giving students the opportunity to read like a scientist!

THE COLLECTION—PURPOSE AND ORGANIZATION

In our conversations with science teachers, we discovered that very few have considered utilizing YAL in the science classroom. The reason for this has been attributed to their lack of knowledge and exposure to YAL as well as instructional strategies they can implement using these texts in their curriculum to develop science concept knowledge. In other words, science teachers are unclear on ways in which they can incorporate YAL into their courses as a medium through which to engage students in dialogues about content to develop scientific literacy. Our book seeks to remedy this by offering educators a collection of practical approaches for teaching science concepts through YAL.

Throughout this collection, authors use the term *secondary* as opposed to middle school or high school; this is intentional as even though some YAL is often associated with specific grade levels, teachers can use these texts across all secondary grade levels. We also refrain from specifying specific reading levels for any of the texts discussed as we have found assigning Lexile and grade levels to be restrictive without adding any benefits. We intentionally leave those types of decisions to our teacher readers who know their students best. Similarly, although the pedagogical approaches offered within the chapters align with current science and literacy standards, we eschewed referencing any specific ones, as lists of standards can become unwieldy in texts and because teachers can determine how activities meet their local requirements.

Each of the chapters presented in this volume is organized correspondingly, with an introductory section, a summary of the text, and then suggested instructional activities before, during, and after reading; furthermore, each chapter includes extension activities that move beyond the text. In many cases, activities build on each other, and in other cases, they exist independently, allowing teachers to pick and choose those that fit their students best.

There is a present-day urgency to refocus students' attention on the natural world as a means to bring awareness to the global issues humanity faces. The chapters in this book have been organized according to the science disciplines their stories address. Each chapter focuses on different subdisciplines of Earth and Life Science. These include the impact of human activity, biology, conservation, biodiversity, genetics, ethics, and bioengineering.

In chapter 1, *Thirsty for Science: Exploring Water Systems, Water Conservation, and Drought through Dry*, Michael DiCicco and Chris Cook explore the many ways *Dry* (Shusterman & Shusterman, 2018) can be read to cover science content such as climate change, water scarcity, water treatment and conservation, water dependency (human-body composition), fire ecology, renewable resources, and weather patterns. Additionally, through the reading of this YAL novel, students have an opportunity to explore ethical, psychological, and philosophical concepts related to climate change and communities in crisis.

Shelly Shaffer and Kathryn Baldwin remind us in chapter 2, *Climate Change Is* A Hot Mess*: The Human Impact on Earth Systems*, that it is critical for young people to have the science and environmental literacy skills to collect and analyze data in order to make informed decisions about Earth. Through the reading of *A Hot Mess: How the Climate Crisis Is Changing Our World* by Fleischer (2021), the approaches and activities in this chapter guide students through the 5E Inquiry cycle with a particular emphasis on improving students' environmental literacy related to the global climate crisis.

Chapters 3 and 4 both spotlight plant science. In chapter 3, *Countering "Plant Apathy": Using Kenneth* Bloom *as a Motivating Tool for Teaching Plant Science to Students*, Katharine Covino and Erin Rehrig demonstrate how *Bloom* by Oppel (2020) can acquaint students with plants and their importance in the ecosystem. The authors offer approaches and activities to help students develop vocabulary and critical literacy skills, foster creativity, conduct inquiry-based experiments on plants, and apply knowledge about the real-world problem of invasive plant species. Julie Baker, Shawn E. Krosnick, and Kelly Moore challenge science educators and their students to connect with introductory biological concepts like binomial nomenclature and cell structure to more advanced topics such as plant chemistry in chapter 4, *Making Botany Magical: Teaching about Plants with* This Poison Heart. References to botanical content throughout the text provide learning opportunities and serve as starting points for activities and discussions on diverse aspects of science literacy. Through *This Poison Heart* (2021) by Kalynn Baylon, students are drawn into a magical world of plants and the people tasked with protecting them.

Chapter 5 shifts to a focus on ecosystems. In *Exploring Nature and the Nature of Scientific Inquiry: Reading* The Evolution of Calpurnia Tate, Amy Palmeri, Emily Pendergrass, and Heather Johnson share approaches and strategies for the exploration of *The Evolution of Calpurnia Tate* by Kelly (2011) in the science classroom. Through the reading and study of this text, students can learn about the nature of scientific discovery—that is how scientists construct, communicate, and critique explanations from evidence. Drawing on the before, during, and after reading strategies described in this chapter, students will have opportunities to follow in Callie's footsteps as they develop and refine their use of a variety of science practices.

The COVID-19 pandemic has created a universally shared point of reference that science educators can use to help teach about disease and immunology. However, since factors including personal trauma and politicized reactions may make it difficult for a teacher to directly use COVID-19 as a teaching tool, stories about previous American pandemics can serve as engaging bridges between past and present experiences of disease in America. Chapter 6 explores these stories. In *Past and Future Plagues as Windows into the Present: Reading* A Death-Struck Year *to Teach about Diseases and Immunity*, David Nurenberg and Ben Lawhorn offer educators approaches for using Lucier's (2014) fictional but historically accurate and highly personal tale of the Influenza Pandemic of 1918 as a tool and reference point for helping students learn concepts such as how diseases spread, how infection happens, the mechanisms of the human immune system, and the function and effects of vaccination. Extension activities are also described for teaching concepts like managing disinformation, exploring the social and cultural aspects of the disease, and honoring the experiences of survivors and first responders during times of crisis.

In chapter 7, *Reading* Ringside, 1925*: Text Support for Teaching Evolution*, Frances Hamilton and Dana Skelley spotlight the ways that the YA novel, *Ringside, 1925* by Bryant (2008) can be read as a complementary text to a science unit on the concept of adaptations from Darwin's theory of evolution. Ways to instruct evolutionary adaptations are shared with the intent that scientific facts can be taught without students feeling their religious beliefs are being challenged.

Janine J. Darragh, Ashley S. Boyd, and Kristina L. Podelnyk share possibilities for using Bodeen's (2010) young adult novel *The Gardener* in a science unit focused on genetics and ethics in chapter 8, *Studying Genetics and Ethics through Young Adult Literature: How* The Gardener *Can Harvest Student Engagement in Biology*. Focusing not only on *how* genetics contribute to traits of living beings but also on the moral dilemmas of whether *should* we manipulate genetics, this chapter demonstrates how a science fiction text can help students relate to and consider not only concepts presented in a science class but also issues impacting the world.

In the final chapter of our collection, Leslie Suters and Kristen Pennycuff Trent offer approaches and activities for using the first novel in *The Hunger Games* (2008) to teach multiple science concepts across life, physical, and earth science curriculum. The authors highlight the multiple ways that this YA novel can be entry points for actively engaging students in three-dimensional instruction focused on disciplinary core ideas, science and engineering practices, and crosscutting concepts. Topics of exploration in chapter 9, *Hungry for More: Exploring, Experimenting, and Engineering with* The Hunger Games, include the examination of genetically modified or engineered organisms, the natural study of ecosystems, survival skills focused on living off the land, medicinal herbs, and water filtration.

REFERENCES

Bayron, K. (2021). *This poison heart.* Bloomsbury.
Boden, S. A. (2010). *The gardener.* Feiwel and Friends.
Bryant, J. (2008). *Ringside, 1925.* Yearling.
Budiansky, S. (2001). The trouble with textbooks. *Prism Online, 10,* 24–27.
Collins, S. (2008). *The hunger games.* Scholastic Press.
Daniels, H., & Zemelman, S. (2004). Out with textbooks, in with learning. *Educational Leadership, 61*(4), 36–41.
Fleischer, J. (2021). *A hot mess: How the climate crisis is changing our world.* Zest Books.
Kelly, J. (2011). *The evolution of Calpurnia Tate.* MacMillan.
Lucier, M. (2014). *A death-struck year.* Houghton Mifflin Harcourt.
Moore, D. W., Alvermann, D. E., & Hinchman, K. A. (2000). *Struggling adolescent readers: A collection of teaching strategies.* International Reading Association.
Oppel, K (2020). *Bloom.* Alfred A. Knopf.
Programme for International Student Assessment. (2015). *PISA for development science framework.* https://www.oecd-ilibrary.org/docserver/9789264305274-6-en.pdf?expires=1661538823&id=id&accname=guest&checksum=547CD038CB34A9578B8E77AA25CBCC6B.
Shusterman, N., & Shusterman, J. (2018). *Dry.* Simon & Schuster.
Wallace, C. S., & Coffey, D. (2016). Science in sync. *Science and Children, 53*(8), 36–41.
Wellington, J., & Osborne, J. F. (2001). *Language and literacy in science education.* Open University Press.

Chapter 1

Thirsty for Science

Exploring Water Systems, Water Conservation, and Drought through Dry

Michael DiCicco and Chris Cook

Water is one of the earth's most precious resources. From sustaining human life, helping provide breathable oxygen, supporting the agricultural needs of the planet, and influencing weather patterns across the globe, water plays a tremendous role in ensuring life continues on earth. As such, understanding the importance of water on earth, the role of water in human biology, and the critical nature of water conservation and sustainability is an essential focus in most secondary education science curriculums across the country. Using the novel, *Dry* (Shusterman & Shusterman, 2018), as a primary text in the science classroom, can be a captivating and meaningful way to engage students in addressing these complex science issues. Though the novel introduces life-or-death scenarios and forces characters to question their own morality and integrity, the story follows two high school students and their journey to survive a natural disaster. The novel introduces opportunities to explore climate change, water scarcity, water treatment and conservation, water dependency (human-body composition), fire ecology, renewable resources, and weather patterns.

The issues of climate change and water scarcity are clearly highlighted throughout the novel: changing climate and weather patterns providing less precipitation to fill reservoirs, local municipalities unwilling to share water resources, and wildfires threatening to destroy communities. The story also serves as a valuable fictional account for students to discuss the potential consequences when these issues are not addressed in real life. In fact, according to a recent NBC news report, almost half of the United States is experiencing drought conditions, with many of the western states experiencing megadrought conditions and increased risk of wildfires. At the time of this writing, CNN reported the country's largest man-made reservoir is at an all-time low level (2022), and CNBC reported many communities are also holding back water releases from dams to keep generating power (2022). While this ensures power for these specific communities, it has a significant impact on the other communities who rely on that water from the dam release. This is exactly the type of scenario that is highlighted in *Dry* and shows the importance of providing opportunities for students to engage in authentic learning activities and make connections to real life.

In addition, the topic of water treatment and conservation are also prevalent throughout the story. The novel takes place on the California coast where efforts to desalinate water (seawater) were unsuccessful and unreliable on a large scale. Is this an issue that could have been avoided? Exploring the importance of water treatment and water conservation can provide a valuable opportunity to examine factors that impact a community's effort to maintain an adequate water supply.

As the lack of water takes its toll on the citizens of Southern California, more and more people turn into "water-zombies"—people who are willing to do anything to get their hands on water. Examining the human body's need for water and how the different body systems could be impacted by dehydration provides a strong connection to the secondary school science curriculum. While the science content is strong throughout the novel, there are also meaningful opportunities to engage students with the themes of crisis management (survival), humanity and morality, and relationships. These themes provide a relevant and current connection to what matters to young adolescents and open the door for rich and engaging discussions about how people's lives can be impacted when we are not learning and acknowledging the science present in our communities.

DRY BY NEIL AND JARROD SHUSTERMAN

What would happen if water was no longer available for human consumption? How would communities respond to a drought of epic proportions? When humanity is faced with a life-or-death scenario, when does the survival of the fittest become more important than morality and human decency? *Dry* brings these questions to life as Southern California is dealing with a catastrophic drought that has depleted all freshwater sources and access to bottled water. When the various water restrictions prove ineffective and the government's desalination efforts fail, people are forced to take matters into their own hands and do anything to survive.

When Alyssa's parents do not return from a trip to find water, she and her brother (Garrett), with the assistance of their neighbor (Kelton), venture out on their bikes in hopes of finding them. Along the way, Henry and Jacqui join the cause, and the crew must learn to trust one another to survive. Ultimately, what starts out as a mission to find Alyssa and Garrett's parents turns into a journey to find safety at the San Gabriel Reservoir—their last hope for water. A whole new world awaits them as they encounter new people, threats to their survival, and an adventure that will change their lives forever. Who can be trusted and what lengths will they go to ensure their safety?

PREPARING STUDENTS TO READ *DRY*

Dry is filled to the brim (pardon the ironic pun) with science content related to water systems, the biology of dehydration, crisis management, wildfires, water conservation, and geography. It also has a number of psychological and philosophical concepts students can explore. Driving the plot is the collapse of the water system in Southern California from a combination of drought brought on by climate change and the political maneuvering of key water sources for communities. Main characters, Alyssa, Kelton, Garrett, Jacqui, and Henry, are thrust into a natural disaster and must navigate a changing world. In addition to simply trying to find water, the group also must navigate the physical dangers of a society in crisis and the psychological and philosophical problems that arise in these situations.

Before reading the text, teachers will want to begin to examine concepts that will be explored in depth during and after reading. We learn there is a drought in many western states affecting an unnamed city in Southern California where our characters live. Students can begin brainstorming how water is used in their daily lives (e.g., water fountains, sinks, garden hoses, pools, fish tanks, showers, watering lawns, water for pets, and toilets), as well as ways water may be wasted (e.g., running toilets, dripping sinks, and watering lawns). As a homework assignment, students can document all the ways they use water in a day or even a week. The class can compile a list of all the ways they have documented their use of water. A version of this can be done at the school site as well with students looking for all the ways water is used in the school building. Alternatively, both of these activities could be done as scavenger hunts created by the teacher. The purpose here is for students to get a feel for how water is integral to their daily lives.

Next, students can explore where this water comes from. We want students to know where the water in the water fountain or from their shower or garden hose comes from. As such, we examine modern water systems including reservoirs, lakes, rivers, aqueducts, and water treatment. Starting from their house, students can research where their water comes from and where their wastewater goes. This can be framed for students as a mystery, a whodunnit of sorts, where they have to find the source of the water entering their house.

Having resources for both public utilities and groundwater supplies in the area will be needed for students to conduct the research. As they move from examining wells or public treatment water treatment facilities, teachers can provide resources on the natural water sources in their area. Students can create a visual map of the water system to their house or community.

Familiarizing Students with the Setting

After the students gain knowledge about water systems, they can then turn their attention to the setting in *Dry*, particularly the Colorado River. The river is key to understanding the novel and many of the scientific concepts discussed in the book. Students should begin to think about its source, flow rate, and how many communities it serves as a water source. The aim is for students to gain an understanding of how many people depend on this river for survival and how it is currently utilized.

In *Dry*, the Colorado River has been dammed by the governors of Arizona and Nevada to save water for their residents which leave the Southern California town without its main water source. Before reading, have students begin exploring historical and current issues with the Colorado River including issues with drought (see table 1.1), laws to help with water conservation, and water rights (see table 1.2). This conversation about current issues with the Colorado River cannot be had without discussing commercial and agricultural water use as well. In particular, students can explore water use by California almond farms during a drought. Media coverage of the California drought has often used California's almond farms as an example of wasteful water consumption. Provided with a number of resources (see table 1.3 for examples) students can debate whether or not California almond farms

Table 1.1. Colorado River Resource Articles

Article	Article Content
Lake Mead plummets to unprecedented low, exposing original 1971 water intake valve (Elam, 2022).	News article on the historic low water level of Lake Mead from the drought.
Wall Street eyes billions in the Colorado River (Howe, 2021).	Focuses on how investors are looking to profit from the drought by trying to buy up water rights.
Water is so low in the Colorado River, feds are holding some back so one dam can keep generating power (Newburger, 2022).	Examines water rights and the importance of water for power in many communities.
'Worst day': Water crisis deepens on California-Oregon line. The Associated Press, NBC News, 2021.	News article on how federal authorities are not releasing water to some farmers and wildlife purposes because of the drought.

Table 1.2. Other Water-Related Articles

Article	Article Content
Columbia's avocado boom shows the hidden costs of "green gold." Megan Janetsky. AlJazeera, 2021.	Explores the costs to rural communities in Columbia as farmers use unsustainable farming practices to keep up with the avocado demand.
Lester Brown: Dust Bowls in Africa and China (video). Lester Brown. Carnegie Council for Ethics in International Affairs, 2015.	Lester Brown is a global environmental activist. In this video, he discusses Dust Bowls beginning to form in Africa and China.
Over half of U.S. in drought as wildfires burn, tornado activity surges. Tim Fitzsimons, NBC News, 2022.	A basic news article reporting on the current state of drought in the United States and the other weather issues related to it.
How water shortages are brewing wars. Sandy Milne, BBC, 2021.	This piece describes the potential future conflict related to water access as more damming and water extraction projects impact others dependent on those water sources.
In town with little water, Coca-Cola is everywhere. So is diabetes (Lopez & Jacobs, 2018).	Reports on the water crisis in San Cristobal de las Casas in Mexico. Despite the town being located in a rainy part of Mexico, they only have clean water two times a week. Coca-Cola has rights to the town's potable water and uses it to bottle their beverages.
The fight to stop Nestle from taking America's water to sell in plastic bottles (Perkins, 2019).	Discusses how Nestle is taking water from federal lands to sell as bottled water.
America's clean water crisis goes far beyond Flint. There's no relief in sight (Worland, 2020).	While the Flint water crisis got a lot of news coverage, this article discusses the many other communities in the United States that are also experiencing issues with accessing clean water.

are wasteful. Students can also similarly explore beverage companies taking water for bottling and the moral, financial, and ecological impacts of damming water for energy use.

Survival in the Desert

Because *Dry* is a novel about survival, students can also explore desert climates and how to survive in the desert. One way for students to do this is to explore the ecosystem and climate of Southern California and what are the natural water sources in addition to the Colorado River. Students can explore native peoples here, such as the Kumeyaay, Cahuilla, Chumash, Tongva, and Serrano, to see how they survived without modern systems for food and

Table 1.3. California Almond Farms Resources

Article	Article Content
Climate change in California is threatening the world's top almond producer. The Associated Press, National Public Radio, 2021.	Explores how drought is affecting almond farms in California.
How almonds became a scapegoat for California's drought (Gonzales, 2015).	Discusses water use for almonds and other crops and explains the reasons for almond farms becoming a favorite target for water conservationists.
California's water problems go way deeper than almonds (Plumer, 2015).	Discusses California's drought and how it is more complex than simply stop producing almonds.
Stop Vilifying Almonds (Hothaus, 2015).	Another piece focusing on the complexity of the water crisis in California.
Almonds swept California Farms. Then the water ran out (Newman, 2021).	Explores the growth of almond farms in California and the effect the drought has had on them.
Almonds: Summary of University of California research on irrigation management for almond trees under drought conditions. University of California, 2022.	An academic site focused on almond tree irrigation. While the text can be a bit dense, it is reader-friendly enough for students to do some research on the practical aspects of almond farming.

water. Exploring how civilizations collected and treated water in communities can make a great cross-curricular connection to history and geography. We can look at clean water sources in the desert and edible plants that can provide hydration. We can begin to discuss the dangers of drinking unclean water, but we want to save the details for when the novel discusses dysentery. As one can imagine in a novel about running out of water, dehydration plays a large role in the story. While we go into much more detail during and after reading, we introduce the idea before the novel, exploring the concept of thirst and why hydration is important to bodily functions.

How Humans Deal with Crises and Natural Disasters

Sociological and psychological issues in how humans deal with crises and natural disasters are explored within the pages of the text. It explicitly discusses major psychological and philosophical concepts. For example, the Call of the Void (p. 123) (the scientific name is High Place Phenomenon) is the feeling to jump one may get when looking over a ledge or a high place. It is not related to suicide. It is part of a group in psychology called intrusive thoughts. In the novel, the concept is discussed by Jacqui when reflecting on her life. Hobbes and Rousseua's debate on the natural state of being (p.

Table 1.4. Example Anticipation Guide

	Strongly Agree	Agree	Disagree	Strongly Disagree
People are naturally selfish.				
If you have everything you need, it is your obligation to help others in need.				
Regardless of your situation, it is your obligation to help others in need.				
It is best to prepare for every scenario you can imagine.				
The ends justify the means.				

291) is also discussed. In this debate, Rousseau believes the state of nature is humans acting on their basic instincts to survive, and this results in a generally peaceful place. For Hobbes, the state of nature is horrible and inevitably results in war. In the novel, the characters discuss which of these will play out in this natural disaster. Machiavelli's concept of ends justifying the means (p. 291) is also discussed. This is the idea that as long as the goal is noble and good, how a person goes about getting there, no matter how awful, is justified. Teachers could use an anticipation guide before reading to explore these concepts and revisit them during and after reading (see table 1.4). The anticipation guide allows for good discussion and an introduction to the text where they see some of these concepts play out early in the novel.

WHILE READING *DRY*

Interactive Notebooks

Dry is separated into six parts, with each part focusing on a specific theme. Teachers can plan during reading discussions and activities around each of these parts. As students read, ask them to keep interactive notebooks sectioned off into the major concepts addressed. Headings for each section include: water systems, water conservation, "water-zombies" (dehydration) (p. 121), crisis management (we include psychological and philosophical elements here), wildfires, and story tracker. Interactive notebooks can be digital, which allows for quick and easy access to maps, sites, and resources. Students can create these notebooks using Google Drive as it allows for flexibility and ease of use. However, OneNote, Evernote, or Keep Notes from Google can work to help students organize their information. For example, students can "pin" locations noted in the book on Google Maps to track their journey and examine the geography of the area. Students can also keep track of research on water systems in one tab, keep track of textual evidence and research on dehydration in another, and have notes and anticipation guides from ethical discussions in another. These can all be done with traditional notebooks as well. Students can use printed maps or create their own to track action in the books.

Interactive notebooks can also help students make connections between science concepts and events in the novel. The science content cannot be isolated from the novel, helping students put context to science concepts and see science's impact on daily life. The notebook also provides a degree of flexibility to modify, add, or delete activities and content based on student needs and class conversation. Below we discuss each part of the book and the activities and content addressed within.

PART 1: THE TAP-OUT

Creating Rain Barrels

When Alyssa, her family, and the community no longer have access to water, it is referred to in the novel as the Tap-Out (p. 4). The novel begins with Alyssa and her family first experiencing life without access to running water. They discover all the ways they depend on water in their everyday life, including with their dog (p. 4). To highlight the challenges Alyssa and her family have to find sources of water and store it, students could create their own rain barrels to discuss water storage and contamination. Creating rain barrels is a fairly simple activity that can be completed with a plastic trash can and plastic tubing. Though the class could easily capitalize on existing downspouts and gutters in the school as a simple way to set one up, allowing students to design and build their own barrels could provide a creative and engaging activity. Requiring students to use recycled or household products (e.g., milk jugs, funnels, garden hoses, and household waste) they have access to provide a more relevant and critical thinking experience. The goal is not to simply build a rain barrel, but think through the different resources available to them and how they could be used to collect water. The students could then explore how the different designs worked and discuss the challenges and successes of water collection.

Charting Locations

Students can devote more time to the study of the Colorado River and Southern California here. While *Dry* is a work of fiction, the locations are real, and the scenarios are realistic. Students can use Google Earth to examine the geology and natural elements of the area and explore the climate of Southern California and look at estimated rainfall, water sources, and ecology. Based on the evidence in the text, students can also chart possible locations of Alyssa's house. In the text, estimated time and distances are given for the beach and nearby areas, but it does not give the specific location. Later students can narrow and revise their estimates as other locations are given and characters travel in the novel (see tables 1.5 and figure 1.1).

Table 1.5.   Example Locations from the Text

Location	Page #
Aliso Creek Canyon	97
Laguna Beach	97
Saddleback Mountain	98
Laguna Canyon Road	108
Pacific Coast Highway	107
Dove Canyon	181
Santiago Creek	256
Santa Ana River	263
Disneyland, Angel Stadium, Honda Center	263
Prado Flood Control Basin	265
Foothill Highway	266
Castaic and Big Bear Lakes	303
San Gabriel Reservoir	328

Figure 1.1.   Example Google Map of Dry Locations.

FEMA

The Federal Emergency Management Agency (FEMA) is mentioned often in the book (e.g., p. 89). While FEMA is dealing with another natural disaster and doesn't have the resources to provide aid during the Tap-Out in the novel, it may be helpful to explore FEMA's purpose, capabilities, and history. The FEMA website (FEMA.gov) provides detailed information about disasters and assistance, emergency management, and resources available to communities. It is helpful for students to understand the different programs and resources that are made available to the community during natural disasters and emergencies. Engaging students in a class discussion focused on Alyssa and her companion's general distrust and reaction to the soldiers at the FEMA camp (p. 244) would be valuable. Why are Alyssa and her friends generally distrusting the soldiers? What would you have done if placed in that situation? What events/emergencies have taken place in our state/community where FEMA has provided support?

Exploration of Ethics

Throughout this part, there are a number of ethical scenarios related to the crisis worth exploring. The decision of Kelton's dad to refuse to give water to a neighbor is one that can provide a rich discussion (pp. 66–71). Kelton's family are known in the neighborhood as "preppers" and end up being well prepared for the Tap-Out with stored food and water, solar panels for energy, and various security systems (p. 21). Knowing this, a neighbor asks Kelton's dad for a bottle of water so he can give it to his wife who is breastfeeding their infant. Kelton's dad refuses, citing they should have been prepared themselves (p. 68). However, he offers them advice on how to get water from the succulents in their own yard (p. 69). Students can explore this ethical dilemma through discussion rather than debate. Debate focuses on having a right answer. Here, have students explore the different positions of the characters and examine whether these positions change or should be changed based on the situation. These are key concepts that come up in different forms throughout the book. Student-led discussions such as Socratic circles or fish bowls work well for fostering discussions for this scenario. Socratic circles are a form of discussion where students research, prepare, and participate in a discussion on a line of inquiry usually related to larger themes and concepts that do not have definitive answers. Fish bowls are discussions where there are two circles of students. In the inner circle students actively discuss a particular topic while the outer circle watches and makes notes about the discussion. Students switch between the circles throughout. Both Socratic seminars and fish bowls provide opportunities for students to fully explore these ethical dilemmas.

Additionally, students can take advice of Kelton's dad and explore how much water can really be extracted from grinding up the succulents (p. 69). Students can research this process that includes which succulents can be ground up and the need for a condenser to extract drinking water from the liquid.

PART 2: THREE DAYS TO ANIMAL

Water-Zombies

Dehydration and how it affects the mind and body is a major topic of this part. The concept of a "water-zombie" (p. 121) is discussed for the first time here. A "water-zombie" in the novel is a person suffering from acute dehydration. Physically they look different. Our characters describe one "water-zombie" as having dry lips, "not just dry, but parched and chapped to the point of bleeding. None of these kids look right. Their skin is thin and almost leprous gray. The corners of their mouths are white with dried spit. And the look in their eyes is almost rabid" (p. 115). It is not just the physical features, "water-zombies" act differently too, with most becoming violent, deranged, and unstable. Students can identify textual evidence like the above and compare and contrast the text with the biology and science of acute/severe dehydration. A simple T-chart can be useful here and students can create visuals of water-zombies on poster paper. (see table 1.6)

Table 1.6. Sample T-Chart

Text from Dry	Symptoms of Dehydration (from various sources)
Lips. "not just dry, but parched and chapped to the point of bleeding" (p. 115).	Dry lips (Mayo Clinic); dry mouth and other mucous membranes (MD Anderson Cancer Center & Cleveland Clinic).
"Skin is thing and almost leprous gray" (p. 115).	Decrease in skin elasticity (MD Anderson Cancer Center).
"The look in their eyes is almost rabid" (p. 115).	Sunken eyes and cheeks (Mayo Clinic).
"His eyes are dark and sullen" (p. 195).	
Attack from water-zombies (pp. 113–120).	Irritability and confusion; extreme thirst (Mayo Clinic); Delirium and confusion (Cleveland clinic).
"he hesitates, gripping the counter and closing his eyes, wincing a bit. He seems weak on his feet" (p. 196).	Fatigue and listlessness (Mayo Clinic); Tiredness, weakness, light-headedness (Cleveland Clinic).

Table 1.7. Water Desalination Resource Articles

Article	Article Content
Why don't we get our drinking water from the ocean by taking the salt out of seawater? Peter Gleick, *Scientific American*, 2008.	Explores why getting drinking water from the ocean is not a viable option.
From seawater to drinking water, with the push of a button. Adam Zewe, *MIT NEWS*, 2022.	Focuses on how researchers have created a portable machine to desalinate seawater.

Desalination Machines

Desalination machines are brought to the beaches to provide some fresh water for the people of the area. These are described as military-style trucks pumping water from the ocean. This leads to a discussion about why we cannot drink seawater and the effect of salt on the body, which connects well to the above discussions about dehydration. Students can explore ways humans have tried to create freshwater from seawater, including desalination machines. In addition, providing supplemental readings (see table 1.7) can help students understand the process, benefits, and challenges associated with water desalination. Further, once students have a general understanding of the process, students can examine if these machines are realistic to use and their practicality in emergency situations, while also discussing why the desalination efforts were not successful in the story.

Exploring Martial Law

Alyssa and Garrett's parents go missing while looking to get freshwater, Kelton's brother is accidentally shot and killed, and Jacqui, a new character, is introduced (p.123). Alyssa and Garrett's parents become lost after a mob erupts at the desalination trucks presumably because there wasn't enough water and people panicked (pp. 108–115). In the chaos, it is clear that some of the trucks were destroyed, people were arrested, and others were killed. Alyssa and company witness the breakdown of society as highways are blocked, cell phones no longer work, and there are riots at trucks attempting to provide water to the community. Events like these lead to martial law being declared. Students can research the concept and history of martial law including its use in US history and around the world (see table 1.8 for example resources). Students can discuss whether the events in *Dry* warrant a declaration of martial law. For context, at the time we are writing this chapter (2022), the war in Ukraine is underway and martial law was declared.

Table 1.8. Example Resources for Exploring Martial Law

Resource	Resource Content
Martial Law. Cornell Law School Legal Information Institute, 2020.	Concise definition of martial law with information on the history of martial law in the United States, who can declare martial law, and example court cases.
6 times martial law was declared, and the constitution suspended in the United States. Military.com, 2022.	A short article discussing the history of martial law in the United States.
What is Martial Law. Nicholas Ferroni. Mental Floss.com. 2012.	A short article explaining what martial law is and examples of it in history.
Ukraine's president declared martial law after Russia's attack. But what is it? Asha C. Gilbert. USA Today, 2022.	Newspaper article discussing how and why Ukraine has implemented martial law.

Exploring Ethical Dilemmas

A number of ethical dilemmas are brought up in this chapter. For example, Alyssa, Kelton, and Garrett attempt to buy a bag of peanuts in a convenience store and are charged $40. Students can discuss the basics of supply and demand and price gouging during crises. Examples from the gas crisis in 2020, Hurricane Katrina, and others can be discussed. An example guiding question may be, is it okay to raise prices for important items when there is a crisis? A small but important change to the question which would alter the conversation would be focusing on necessary items. After watching Michael Munger's video on price gouging or reading parts of Matt Zwolinksi's paper "The Ethics of Price Gouging," students are presented with arguments for why price gouging is a good thing during a crisis. Further, Alyssa steals water from Kelton's family and gives water to the adults during their homeowners' meeting in an attempt to help families in need. She watches as the adults struggle to share the water and debate who gets water and who doesn't. The meeting escalates quickly with one adult telling Alyssa to get out before it becomes violent (pp. 157–160). This situation mirrors what is happening on a larger scale in Southern California.

PART 3: THE CHASM BETWEEN

Dysentery

Part 3 focuses on water quality and the dangers of drinking contaminated water. The characters meet Alyssa's Uncle Herb in a housing development in Dove Canyon (p. 194). We learn Uncle Herb, his girlfriend, and the rest of the community have been able to sustain themselves longer than the rest of Southern California because of access to an old water tank. The tank hasn't been used in years but had water in it. However, many of the town's people have fallen ill, including Alyssa's Uncle Herb. He is described as "pallid, and his face seems to sag, like his skin has grown tired of clinging to the bone. His eyes are dark and a little sunken. He looks like a drug addict, but I know that's not it" (pp. 195–196). It becomes clear that Uncle Herb, his girlfriend, and the town are suffering from dysentery. Students can explore the bacteria, symptoms, and cure. Connections to the Oregon Trail game can be made and students can explore why dysentery is less of an issue today than it was back in the time of Louis and Clark. There are multiple mentions of characteristics of dehydration in this part including checking skin elasticity for dehydration levels and a burst of energy at the end of life similar to "water-zombies" in the part above (see table 1.6). If a T-chart wasn't completed in the last section, it can be done here.

Water Conservation Laws

Water conservation can also be explored and students can identify ways to conserve water in their own homes and communities. In the novel, a "frivolous use initiative" (p. 190) is highlighted, which is meant to curb the non-essential use of water, such as watering a lawn or the use of decorative fountains. While this particular initiative is not implemented in real life, there are similar current initiatives enacted throughout the United States, particularly in many western states. The Environmental Protection Agency has been working to reduce water consumption and encourage communities and citizens to enact water conservation laws since the early 2000s. Students can research these initiatives and explore what changes states and communities have made to conserve water. Students can work in small teams and create a graphic organizer highlighting the initiative found, where it takes place, and the conservation efforts called for with the initiative. Teams can then share their findings with the class. For example, students might find Nevada's ban on watering ornamental grass or Los Angeles' watering restrictions. The point of the activity is not to develop an exhaustive list of conservation efforts, but rather to see the different things being done to help address the issue of water availability. Students can then make connections to their own lives and identify ways to conserve water in their own communities.

Moral Implications

In Uncle Herb's neighborhood, we meet a new character Henry, who has been trading bottled water for gifts and necessities (pp. 185–188). Henry has a stash of hundreds of bottles of water. His community is suffering from a lack of clean water. Instead of charging money, he has asked people to give up prized possessions, so that when the crisis is over Henry will be rich. He states, "If I've learned anything in my studies, it's that the greatest investors capitalize in times of crisis. And though at first it may sound cold, it's the giant's duty to continue to stand tall and generate profit, which will lead to spending, and ultimately stimulate the economy for the greater good" (p. 186). This is reminiscent of the bag of peanuts scene in the last part. Students can add to their discussion from the last part on the moral implications of Henry's actions. A simple four-corner whole-class discussions where students answer guiding questions by moving to the corner of the room with their answer (i.e., strongly agree, agree, disagree, and strongly disagree) is a fun way to have this discussion.

Aqueducts

Mapping the character's journey continues here with the Prado Control Basin and the Angeles National forest being specifically mentioned (p. 265). Aqueducts feature heavily in their journey during this part of the novel. Students may be unfamiliar with these and their purposes. Aqueducts in modern times are used to move water from its original source to distribution sources. While the goal of the characters is to get to the bug out, these aqueduct paths foreshadow their journey to a source of water that will later save them. In an effort to highlight and understand the challenges of designing an aqueduct, students can create aqueduct models in class to show how they work. For this activity, students would use a variety of craft materials (e.g., cardboard, craft sticks, and aluminum foil) and design an aqueduct structure that would move from point A to point B while navigating around different obstacles (e.g., gravity, elevation changes, and rock formations). Once designed, the models can be tested by having students see how much of a twenty-ounce bottle of water makes it from point A to point B. Students should then reflect on the design and propose changes for a rebuild.

PART 4: BUG OUT AND PART 5: HELL AND HIGH WATER

Topographic Mapping

Parts 4 and 5 include the climax of the novel and focus on wildfires and dehydration. The characters finally reach the bug out, a hide-out stocked with emergency provisions used in emergencies, such as natural disasters, only to find it has been raided and doesn't have any supplies (p. 307). Students can discuss what the term "bug out" means in military parlance and how they work. Students can attempt to locate where Kelton's family bug out would be on a map based on their tracking of the action in Castaic Lake and the San Gabriel Reservoir. Part 5 mentions topographic maps so that this may provide an opportunity to explore reading those types of maps and possibly find or create one for their own community. Topographic maps show elevation changes in the landscape in addition to other natural features such as lakes, roads, and terrain. The US Geological Survey website (usgs.gov) has a number of features including information on how to read topographic maps and maps for the entire United States. They include maps of Castaic Lake, San Gabriel Reservoir, and other locations in the novel. Additionally, at the time of writing, the site has a feature for mapping, tracking, and learning from wildfires.

The Science of Wildfires

From property destruction, smoke inhalation, and the threat of death, wildfires go from being a looming peripheral threat in Part 3 (p. 293) to an immediate danger in this part of the novel (pp. 347–361). Students can explore the science of wildfires including what conditions are needed for them to spread, how they start, and how fast they move. Exploring the concepts of what makes the fires travel quickly and why they are faster going uphill than downhill can be interesting to students. Unfortunately, wildfires have been a more common occurrence in the United States over the past twenty-five years. Students can research different fires and the impact they have on their communities (e.g., Gatlinburg, Tennessee in 2016, Bay Area Fire in California in 2020, and Dixie Fire in California in 2021). A natural conclusion to the activity and class discussion is how to stop and prevent wildfires. Students can explore the methods firefighters currently use to put out wildfires, as well as make connections to the recent wildfires in California and in the western United States.

PART 6: A NEW NORMAL

Water Conservation

This short section concludes the novel. As the title of this part suggests, Alyssa and company are back with their families. They are safe and living a new normal where water is scarce and water conservation measures have to be taken. The simple act of taking a shower or giving a pet water is now different. Water is turned on only two days a week, so Alyssa's family must find ways to conserve (p. 377). This is a great opportunity to continue the discussion about water conservation started in Part 3 and examine what can be done by individuals, communities, and governing bodies to conserve water. Students can develop specific plans and proposals for how they can conserve water as both an individual/family, as well as part of the specific classroom community. The classroom community plans can be shared with the entire class and then one of the class plans can be selected to be put into action. Students would have firsthand experience to measure the progress of water conservation in their own classroom and have relevant examples to discuss the benefits and challenges associated with water conservation efforts.

Debate

The governor in Arizona is removed from office and charged with crimes for blocking off the Colorado River and causing Alyssa's community to lose its water source (p. 382). Students can debate the governor's actions of cutting off a community's water source to save their own people. Did the governor do the right thing by cutting off the river to save his own community? Using their knowledge of water systems, students could create possible alternative solutions to the governor's dilemma. The text mentions the Tap-Out disaster had the most deaths of any non-war event (p. 376). The California wildfires of 2015–2018 are considered one of the deadliest US-based natural disasters. Students can explore what caused these fires, what made them so deadly, and make connections to how they are described in the novel. Students can also explore recent and historical natural disasters such as the Fukushima nuclear disaster caused by an earthquake and tsunami in 2011, the Sichuan earthquake of 2008, Australian wildfires of 2019–2020, and Hurricane Maria in 2017, which was the deadliest US-based natural disaster in the last 100 years.

AFTER READING *DRY*

Children's Book Project

Children's books can be a powerful way to introduce meaningful issues to young readers. With access to clean, freshwater being an issue for millions of people across the globe, students could write and illustrate children's books that highlight the various issues impacting water quality and availability addressed in *Dry*. This could be an enjoyable and creative way to capture the key lessons learned in the novel, as well as develop resources that could be shared with others.

Service-Learning Project

In an effort to answer what could be done to avoid a local "Tap-Out," students can also research the conservation efforts in their local area. A service-learning project could explore how water is used in their local community and what is being done, if anything, to conserve water. A natural connection is for students to explore their local region's water system. They can investigate the source(s) of water in their area and how it is collected, held, and treated. Reservoirs, aqueducts, and other concepts from earlier can be explored here. They can investigate who owns the water rights. Is it possible for the local water source to be "turned off" for political or other reasons? Students can also explore what process is used for treating the water in their area. Students can explore what laws are in place or being proposed related to water. After researching their own water system, students can suggest ways to improve the system or awareness of water conservation.

How Water Is Used

Students can explore how water misuse can impact communities and result in similar outcomes to the Tap-Out in *Dry*. This discussion can lead to how water is used in agriculture. For example, students can investigate and debate issues related to agriculture such as California almonds during the drought and avocado farms in South America. Similarly, students can investigate clean water access across the United States and the world. The crisis in Flint, Michigan, along with other US cities with clean water issues including Pittsburgh, Pennsylvania; the Navajo Nation, East Chicago, Illinois; Airway Heights, Washington; Brady, Texas; Baltimore, Maryland; Milwaukee, Wisconsin; and Las Vegas, Nevada can help students understand that clean water access continues to be an issue even today.

EXTENSION ACTIVITIES

Service-Learning Activities—Children's Literacy and Water Conservation

Building off the children's book activity as part of the after-reading strategies, the class could partner with a local elementary school and share their books with the younger students. In addition to sharing their stories and celebrating literacy, the focus could also be on how to impact water availability and conservation in their own community. A service-learning project could explore how water is used in their local community and what is being done, if anything, to conserve water. Students can explore what laws are in place or being proposed related to water. After researching their own water system, students can suggest ways to improve the system or awareness of water conservation. Students could brainstorm together and focus on changes they could make to have an impact on their immediate community. These brainstorm sessions could potentially lead to specific projects at their individual schools (e.g., water collection and information sharing) or also be part of another service-learning activity where the secondary school students help the elementary school host a "Water Fair" for the community.

Surviving the Wilderness

With survival being a key theme of the novel, providing an outdoor adventure experience would be a great experiential learning opportunity for students, as well as a great way to build community among the class. While this knowledge may not be something most teachers possess, many community and commercial partners are likely to have resources to help in the planning or scheduling of this activity (e.g., Outward Bound, REI, YMCA, and environmental education programs). This could be scheduled as an on-site half-day activity where the outdoor educator offers hands-on demonstrations or an overnight camping adventure where students are able to practice the skills firsthand in a controlled environment. Students could learn how to set up camp, access fresh water, find food sources, and handle different emergency situations that arise (e.g., first aid, plant identification, and orienteering). While the cost of such an experience may pose a financial burden, there are grants and philanthropic organizations focused on providing these experiences for students.

Classroom/Community Garden

Water is essential in creating a garden and sustaining life. Creating a classroom or community garden where students take the lead role in establishing and maintaining the space provides a meaningful way to build community and illustrate the value of water for both life and food sources. Students could also focus on filling their gardens with plants native to their community. This could also be a great opportunity to enhance the school community by having the harvest of the garden be sold at a school event and profits of the garden be donated to a water conservation or local food pantry.

Advocating for Water Conservation

Students can work together to create a guide, presentation, or signage (pamphlets, posters) to convince their community or government council to take water conservation seriously. It can include an overview of the water system, including identifying main water sources and how it's processed and treated for the community. Students can note potential threats to the water system including natural (i.e., drought) and political (i.e., shutting off water sources), suggest water conservation solutions, and explain why they would be effective for the community.

CONCLUSION

It is no secret that the world is facing a global climate crisis, and this crisis is impacting billions of people across the globe. In the United States alone, significant drought, risk of wildfires, and depleted water supplies are forcing many communities to enact water-use policies and restrictions. With the changing weather patterns and decreased access to clean water, more and more people will be forced to make changes to sustain life for future generations. *Dry* offers an exciting journey through crisis management as the characters navigate complex challenges to stay alive. Exploring this novel with secondary students would provide tremendous opportunities to engage in rich discussion, make authentic connections to real-life scenarios, and think critically to solve complex problems—all essential attributes for meaningful learning. These learning experiences could potentially make a difference in communities around the world. Engaging students in these discussions now may be the difference that is needed in the future.

REFERENCES

Elam, S. (2022, April 29). Lake Mead plummets to unprecedented low, exposing original 1971 water intake valve. *CNN.* https://www.cnn.com/2022/04/27/us/water-intake-exposed-lake-mead-drought-climate/index.html.

Fitzsimons, T. (2022, May 9). Over half of U.S. in drought as wildfires burn, tornado activity surges. *NBC News.* https://www.nbcnews.com/news/weather/half-us-drought-wildfires-burn-tornado-activity-surges-rcna27976.

Gonzales, R. (2015, April 16). How almonds became a scapegoat for California's drought. *National Public Radio.* https://www.npr.org/sections/thesalt/2015/04/16/399958203/how-almonds-became-a-scapegoat-for-californias-drought#:~:text=How%20Almonds%20Became%20A%20Scapegoat%20For%20California's%20Drought%20%3A%20The%20Salt%20%3A%20NPR&t

ext=Press-,How%20Almonds%20Became%20A%20Scapegoat%20For%20California's%20Drought%20%3A%20The%20Salt,water%20to%20grow%20one%20almond.

Holthaus, E. (2015, April 17). Stop vilifying almonds. *Slate.* https://slate.com/business/2015/04/almonds-in-california-they-use-up-a-lot-of-water-but-they-deserve-a-place-in-californias-future.html.

Howe, B. R. (2021, January 3). Wall street eyes billions in the Colorado's water. *The New York Times.* https://www.nytimes.com/2021/01/03/business/colorado-river-water-rights.html.

Lopez, O., & Jacobs, A. (2018, July 14). In town with little water, Coca-Cola is everywhere. So is diabetes. *The New York Times.* https://www.nytimes.com/2018/07/14/world/americas/mexico-coca-cola-diabetes.html.

Newburger, E. (2022, May 3). Water is so low in the Colorado river, feds are holding some back so one dam can keep generating power. *CNBC.* https://www.cnbc.com/2022/05/03/lake-powell-glen-canyon-dam-water-release-delayed-due-to-drought.html.

Newman, J. (2021, July 5). Almonds swept California farms. Then the water ran out. *Wall Street Journal.* https://www.wsj.com/articles/almonds-swept-california-farms-then-the-water-ran-out-11625490000.

Perkins, T. (2019). The fight to stop Nestle from taking America's water to sell in plastic bottles. *The Guardian.* https://www.theguardian.com/environment/2019/oct/29/the-fight-over-water-how-nestle-dries-up-us-creeks-to-sell-water-in-plastic-bottles#:~:text=Nestl%C3%A9%20is%20now%20fighting%20a,to%20speak%20to%20the%20controversy.

Plumer, B. (2015, April 14). California's water problems go way deeper than almonds. *Vox.* https://www.vox.com/2015/4/14/8407155/almonds-california-drought-water.

Shusterman, N., & Shusterman, J. (2018). *Dry.* Simon & Schuster.

Worland, J. (2020, February 20). America's clean water crisis goes far beyond Flint. There's no relief in sight. *Time.* https://time.com/longform/clean-water-access-united-states/.

Chapter 2

Climate Change is *A Hot Mess*
The Human Impact on Earth Systems
Shelly Shaffer and Kathryn Baldwin

Environmental changes are visible across the United States—and around the world. In *A Hot Mess: How the Climate Crisis Is Changing Our World*, Fleischer (2021) shares how tornado, fire, hurricane season, extreme drought, and severe flooding affect human lives and explains how these events have become more severe due to human impact on the environment. Written for adolescents, Fleischer uses scientific facts and data throughout the book to illustrate how these changes to the environment are not due to the weather cycle, rather, they are the result of more serious climate change and reflect evidence of a global climate crisis.

A Hot Mess discusses how increased global surface temperatures, melting ice, and rising sea levels are at the root of the environmental crisis today. Human impact on the Earth has caused the entire planet to change—and heat up. Fleischer (2021) identifies an environmental crisis (a mess) that is close to reaching its tipping point; we are running out of time to "fix things" (e.g., greenhouse gases, carbon emissions, global warming, and melting icecaps) (p. 28). He writes,

> Some people talk about the climate crisis in terms of saving Earth, but the truth is the planet itself will survive. It was around for billions of years before we came along, and it's going to be around long after us. A better way to talk about addressing climate change is that it means trying to save humanity and the other life on Earth. (p. 28)

This chapter focuses on the *mess* that humans have wreaked on the environment, challenging readers to explore topics such as wildfires, rising sea levels, drought, and deforestation in critical ways. Through guided discussion and activities, we share ideas for integrating science with literacy in concrete ways, including data literacy and visual literacy through the reading of *A Hot Mess*. It is critical that young people have scientific literacy (i.e., data analysis, recognizing patterns, and synthesizing research) and environmental literacy (i.e., critical thinking, analysis, and making connections) to make informed decisions about the Earth.

Throughout the chapter we utilize the 5E Learning Cycle Model (Bybee, 2015), often referred to as the 5Es, as a way to organize activities and lead students through an inquiry unit focused on climate change and human impact on the environment. The cycle's five steps, *Engage, Explore, Explain, Elaborate,* and *Evaluate,* guide students through the cycle of scientific inquiry and is a common method used in STEM classrooms. The *Engage* stage hooks students' interest, elicits prior knowledge, and prompts questioning. The *Explore* stage focuses on data collection and analysis, often through investigations that focus on specific parts of the topic. The *Explain* stage, in the middle of the cycle, asks students to explain the concepts they have learned about during both the *Engage* and *Explore* stages. *Elaborate* stage extends conceptual understanding and/or answers questions from earlier stages. Lastly, in the *Evaluate* stage, teachers assess student learning, and students assess their own learning of the unit objectives. With this in mind, the activities throughout this chapter feature activities using one (or more) of the 5E Stages, organized chronologically to lead students through the 5E Cycle.

A HOT MESS: HOW THE CLIMATE CRISIS IS CHANGING OUR WORLD BY JEFF FLEISCHER

A Hot Mess: How the Climate Crisis Is Changing Our World by Fleischer (2021) is a nonfiction young adult book that addresses the global climate crisis. Fleischer uses stories and anecdotes to illustrate and make visible the scientific data and information being shared.

The data in the seven chapters is shared in multiple modes—graphs, photos, and text boxes. This use of multiple formats for data allows readers to examine tangible evidence of environmental crises and incorporates data and visual literacy. Each chapter focuses on a different way the Earth (and life) has been impacted by humanity. Chapter 1 addresses the basics of climate change. Chapter 2 centers on extreme weather across the world as Fleischer shares how hotter summers, wetter weather, and severe storms have become more devastating than in the past. Chapter 3 shares how drought and deforestation have contributed to extreme fire seasons. Chapter 4 focuses on melting glaciers and polar ice caps, leading to higher sea levels. Life around the world, including human, animal, and plant life, is addressed in chapter 5. The way climate change has impacted human behavior is addressed in chapter 6. Chapter 7 challenges readers to consider their own actions toward undoing the damage caused by the climate crisis. Fleischer ends the book with a call to action: "there's a chance to rise to the occasion. If young people—like many of you reading this–get educated and active, they can literally save the world" (p. 175).

PREPARING STUDENTS TO READ *A HOT MESS: HOW THE CLIMATE CRISIS IS CHANGING OUR WORLD*

Eliciting prior knowledge is a key characteristic of the *Engage* stage of science inquiry. Pre-reading activities should tap into students' interests and prior knowledge; the before-reading strategies we share develop the Engage stage of the 5E Learning Cycle Model (Bybee, 2015). Prior to reading *A Hot Mess*, teachers engage students with the concept of climate change. We suggest the following activities that spark students' curiosity and build on prior knowledge. Using the gallery walk and brainstorming activities that follow evokes prior knowledge of weather, climate, and global climate change. Additionally, students are given opportunities to generate their own questions about global climate change during this stage, which will lead them into the *Explore* stage that occurs during reading.

Engaging with the Climate Crisis through a Gallery Walk

A Hot Mess highlights the impact of the global climate crisis. To engage students in considering the various forms of human impact on the global climate crisis, seven to ten photographs that illustrate four main categories of climate change impact (wildfires, extreme weather, and flooding) and other impacts of global warming such as hotter temperatures and increased drought, declining water supplies, reduced agriculture yield, warming oceans, melting glaciers, and sea-level rise, should be printed out and posted throughout the classroom (see figure 2.1 for sample photographs).

Each photo should be numbered, so that as students walk around the classroom to view each, they can note their observations for each specific image. They should use a graphic organizer to think critically about each image (see table 2.1). The *Gallery Walk Graphic Organizer* can help students organize their thoughts about the photographs, as they are required to include a description of the picture, an analysis of how these images might connect to one another, predictions about the unit and science concepts, and questions that can be used to guide them during reading as they learn more about the climate crisis.

After students have rotated through the images in the gallery walk, they can share their observations in small teams and note any differences, adding

Figure 2.1. Sample Photographs of Climate Change Impacts. *Source*: Photographs taken by authors.

to their own graphic organizer as needed. Once a set team discussion time has expired, the teacher can walk around the room, stopping by each image, prompting students to share their analysis, predictions, and questions for each visual.

Brainstorm and Word Cloud

A Hot Mess begins with a discussion of the difference between climate and weather (pp. 16–18). To prepare students to consider these words in the context of the book, they are presented with the words: "climate" and "weather."

Science notebooks are a common practice in science classrooms and are used to provide a space to record and reflect inquiry-based observations (Fulton & Campbell, 2014). Using their science notebooks, students individually brainstorm what they know about these words by creating a Venn diagram (Fulton & Campbell, 2014) and drawing or writing their thoughts about these two words; they should be encouraged to include questions and to consider how these words compare. Then, students share their definitions aloud while the teacher writes on the whiteboard or projector, being sure to include all ideas. The class can come up with a working definition for each term based on the discussion, and each student can write the working definition in their notebooks. This provides a hint for students that the upcoming unit connects to these concepts, and students begin predicting that the photographs they examined during the first activity might also relate to climate and weather, making additional connections.

Climate Change is A Hot Mess

Table 2.1. Gallery Walk Graphic Organizer

Image Description	Analysis	Predictions	Questions
• Describe what is being depicted in these images. • Describe what is happening.	• In what ways might these images relate to one another? • In what ways are the images the same? Different? • List what you already know about this image.	• Make a prediction about what the unit might be about based on these pictures. • In what ways does each photograph connect to science?	• What questions do you have about this photograph? • What would you like to know more about? • What kinds of data will I need to help me understand this photograph?
Example Responses			
Haboob Photo This image depicts a dust storm. It looks like it is rolling in. There is a clear line of dust. It looks like a desert climate.	These images are both showing nature and some kind of disaster. Each depicts a different environment: desert and forest. Both show air pollution, either from the dust or the smoke, so there are some similarities in the images themselves. I wonder if the forest fire was caused naturally or by humans. I think the haboob was naturally occurring. I have never seen a haboob, but I have heard about dust storms before. I have lived near forest fires, so I am familiar with this.	• The unit might be about weather or desert climates.	• Why is there a dust storm? • What causes the dust to roll in that way? • What causes dust storms?
Fire Photo This image shows a forest fire. We can clearly see smoke, fire in the bottom of the trees, and some vehicles in the foreground.		• I think the unit might be about dry weather/climates. • Both of these photos show the result of dryness in the environment. • The unit might be about the environment or desert environments in particular.	• Was the fire caused by humans? • What causes so many forest fires during certain times of the year? • Why does it seem like there are more forest fires taking place lately than in the past?

Note: Photographs of Global Climate Crisis (completed by authors).

Following the initial discussion, students can brainstorm additional words related to climate and weather, such as global warming, rain, heat, or fires, by adding these words below their Venn diagram in their science notebooks. When the class has exhausted all of their ideas and the whiteboard or projector is filled, students transition into a creative activity. In this activity, students create an artifact (i.e., poem, collage, and word cloud) that illustrates the connections between these concepts and their meaning (see figure 2.2 for an example project).

WHILE READING *A HOT MESS: HOW THE CLIMATE CRISIS IS CHANGING OUR WORLD*

During reading, students shift to the *Explore* step of the 5E Learning Cycle (Bybee, 2015). During this stage, students begin to build knowledge about the topic by collecting and analyzing data about climate change.

Figure 2.2. Example Project. *Source:* Created by authors.

Finding Facts and Asking Questions

Students need to be able to strategically read a variety of science texts (Mawyer & Johnson, 2017), including popular texts, textbooks, and primary scientific literature in order to develop their scientific literacy skills. Teaching students to read like a scientist means that teachers should incorporate disciplinary and content-area literacy into their teaching (Mawyer & Johnson, 2017; Zygouris-Coe, 2015). Students who are able to read like scientists engage with texts, including being able to synthesize ideas in text, data, and visuals; determine a main idea and provide an accurate summary; create a schema by connecting prior knowledge to new information; distinguish opinions from facts; and use critical thinking to question bias. To promote this level of engagement with *A Hot Mess*, teachers can lead students through a guided exploration of the text. Students can use their science notebooks to record responses to some of the following guided reading prompts as they read:

Introduction (pp. 6–13)

- What did you learn about Tuvalu from the author's story (pp. 6–9)?
- How is Tuvalu being impacted by environmental change (pp. 6–9)?
- According to the author, how is the focus of this book different than other climate books (p. 10)?
- What do you predict the book will be about based on this introduction?

Chapter 1: Climate Change 101 (pp. 14–43)

- Find evidence presented by the author proving that climate change is real (pp. 15–16).
- Explain the difference between "climate" and "weather" (pp. 16–18).
- How do greenhouse gases impact climate change (pp. 18–21)?
- The cause for the changing climate is different today than it was in the past. Please explain (p. 21).
- There are several factors attributed to the current climate crisis. What are a few the author shares (pp. 21–28)?
- Analyze the table on pages 26–27. (Refer to "Data Patterns" activity.)
- What proof have scientists found about climate change (pp. 28–37)?
- Take specific note here of one current study and share facts found by that study (pp. 28–37).
- On the other side of the environmental science debate are climate deniers. What are the reasons some people might ignore the science about climate change (pp. 37–43)?
- Take a look at the charts on pages 40–41. What conclusions can you draw about the sources of carbon emissions?

Chapter 2: A Change in the Weather (pp. 44–63)

- The author begins this chapter by identifying some changes in global weather. What are two of these examples (pp. 44–46)?
- On pages 47–49, the author writes about heat waves. How has your community been impacted by heat waves? How do the hottest summer temperatures in your area compare to the hottest temperatures in the past?
- On page 50, the author claims that understanding the difference between weather and climate (remember the Explore activity and the class definition) is key. How has the weather in your area changed? How does this connect to the climate crisis?
- On pages 53–63, the author discusses increased flooding and severe weather. Using critical literacy, consider how flooding and severe weather have impacted different groups of people? Which groups are more likely (or less likely) to recover from these events quickly?
- The author shares a bar graph on page 61. This graph shares how pollution, climate change, and environmental issues disproportionately impact people of color. Why do you think this is the case?

Chapter 3: Fire and Other Alarms (pp. 64–87)

- What is a drought (pp. 70–74)? What are the reasons drought occurs?
- On pages 64–70, the author shares an example from the California drought. How has this drought impacted the lives of people living in that area?
- How are droughts connected to human behavior (p. 74)?
- How have droughts impacted humans? What are humans doing to curb those impacts (pp. 74–78)?

Chapter 4: The Tide Is High (pp. 88–109)

- Using data provided from pages 89–92, create a visual representation of these data. NOTE: You don't have to use all of the data and may choose to focus on a specific element.
- The author identifies four island nations that are particularly vulnerable to sea-level rise (pp. 92–93). How has sea-level rise impacted people's lives on these islands?
- The visual of the polar ice caps on page 98 connects to the text describing ice melt on pages 94–101. How has ice melt accelerated in the past twenty years? How does this impact the climate?
- The "Other Rising Concerns" section (pp. 103–106) links melting ice caps to other destructive impacts on the Earth. Describe at least one of those impacts, linking to prior chapters in the text. How will humans be impacted if sea levels rise by one foot? Two feet? Three feet (pp. 106–109)?

Chapter 5: Life during Warming Time (pp. 110–129)

- In this chapter, the author begins to discuss how life on Earth is impacted by climate change. What sorts of life are most impacted (pp. 110–129)?
- Choose one species that has been impacted by climate change and describe how either wildfire, extreme weather, or flooding have impacted this lifeform.
- The author claims that mosquitos and ticks are thriving in the warmer and more humid weather caused by climate change (pp. 127–129). Why do you think that these parasitic creatures are thriving while so many others are suffering?

Chapter 6: A Changing Social Climate (pp. 130–151)

- The author introduces three terms to describe people who are displaced as a result of climate change: internally displaced persons, climate migrants, and climate refugees (pp. 132–138). How would you describe the difference between these three categories?
- Where does the author suggest that internally displaced people in the United States could potentially find refuge? Do you think this solution makes sense (p. 133 textbox)?
- The author shares several examples of internally displaced people in the United States. According to pages 134–138, the author points out that poor and marginalized people are impacted more often than others. Why is it so difficult for poor and marginalized people to move?
- Examine the map on pages 138–139. This map shares locations where climate change is occurring and thus where many climate refugees throughout the world may come from. Where do you notice droughts? Where are people most impacted by severe weather? Where in the world is flooding most likely to happen? Are there certain parts of the world that seem to be more at risk than others? You may also refer to pages 138–142 for additional information in the text.
- Pages 142–144 share information about how climate change has impacted Central America. Describe the connection between warming weather and migration from this area.
- Fleischer connects several events to climate change on pages 145–151. Identify two causes and two effects from these examples.

Chapter 7: So, What Can We Do about It? (pp. 152–175)

- To begin the chapter, the author writes, "how people use common resources such as water, air, and land affects how others can use those resources . . . What's happening to the climate is happening to everyone" (p. 152). Based

on what you've learned throughout this book, list ways that climate change is personally impacting you or your family.
- In chapter 7 (pp. 152–175), the author shares examples of actions we can take to help stop or reverse the impact of climate change. Describe at least two suggestions and connect each to sea-level rise, drought, or severe weather.
- The author concludes, stating

> Thanks to human-made climate change, the world is not only a lot different than it was before the Industrial Revolution, but a lot different than it was just a generation ago. The problem has been more than a century in the making, but young people are the ones who will live through the worst of the crisis. . . . They've inherited a mess [emphasis added], and it's going to take a lot of work by a lot of people to clean it up. That isn't fair, and it's not their fault. (p. 174)

Respond to this quote, connecting with what you've learned from the text and thinking critically about what this means for your future.

Students can also consider these more general questions about the text, which include connections to prior knowledge and exploration:

- What is being discussed?
- How does the current topic relate to my prior knowledge?
- How does the current topic relate to other chapters I have already read?
- What questions do I have about the topic?
- How does the topic relate to a specific environmental science concept?
- What would I like to know more about?
- What additional data would help me to understand the topic more clearly?

Building on the questions students generated during the pre-reading gallery walk activity, students may also seek to answer their previous questions. For example, in the pre-reading activity, we generated questions relating to the haboob and wildfire photographs: *Why is there a dust storm? What causes the dust to roll in that way? What causes dust storms? Was the fire caused by humans? What causes so many forest fires during certain times of the year?* and *Why does it seem like there are more forest fires taking place lately than in the past?* While reading the text, we could find insight into our previous questions (see chapter 2: "A Change in the Weather," pp. 44–63 and chapter 3: "Fire and Other Alarms," pp. 64–87). Students can record evidence to answer these questions, including facts and data shared by the author. At this point in their science exploration, students are still Exploring and learning about climate science using the provided text, *A Hot Mess*. Additional research and elaboration will take place after

Data Literacy

Data produced by scientific investigations must be analyzed in order to understand the meaning (NSTA, n.d.). Data is often presented in a form that can reveal patterns and relationships (i.e., line graphs and bar graphs), and being able to analyze scientific data is a skill identified within science literacy. In fact, being able to integrate quantitative or technical information in both words and visuals is a science literacy skill that students need to be able to master.

Comparing Data

In this activity, students analyze three different data sources from chapter 1, looking at each graph separately and then comparing the graphs to one another. Students write their answers in their science notebooks, clearly labeling the response and citing the graph they are responding to.

To begin, have students examine the "Global Mean Surface Temperature" graph (p. 16) and respond to the following prompts using their science notebooks: *(1) What do you notice about this graph? (2) What questions do you have?* and *(3) Imagine that you were only given one of the two lines, what data might be lost and how might this lead to a misinterpretation of the graph?* For example, on the "Global Mean Surface Temperature" graph (p. 16) if the author only shared the five-year mean, surface temperature extremes (highs and lows) might not be evident from the graph.

Students encounter another line graph further in chapter 1 (p. 31), where the author presents data about atmospheric CO_2 and CO_2 emissions. Students should contemplate and answer the same questions about this new graph (p. 31) in their notebooks. On page 31, an example of data that might be lost or misinterpreted if the graph were simplified is the recognition of similar patterns to changes in CO_2 emissions and atmospheric CO_2.

Next, in small teams (four to five students), students should discuss their responses to the questions about each graph. To check for student understanding, have the small teams share out with the class and discuss any misconceptions. One possible misconception to watch for is conflating weather and climate. Teachers can refer back to the author's discussion of "weather" and "climate" on pages 16–18 and also the pre-reading activity exploration that students participated in.

Since students have now analyzed each of these graphs in isolation, prompt them to place the two graphs next to one another. Pose the following prompts and ask students to first respond in their science notebooks and then to discuss in small teams: *(1) Describe any similarities between these two graphs, (2) Describe any differences between these two graphs, (3) Make an evidence-based claim of changes to atmospheric CO_2, CO_2 emissions, and global mean temperature changes, and (4) Considering these two examples of line graphs, discuss when it is best to use a line graph to represent data.* These prompts guide students in making comparisons between the graphs and synthesizing their learning about line graphs. Students should note similar patterns, as atmospheric CO_2, CO_2 emissions, and global mean temperature changes all increase over time. Students should be interpreting the data and making an evidence-based claim, based on the information presented in the graphs.

Data Patterns

Students can examine the "Where Carbon Comes From" textbox (pp. 26–27), to further develop scientific literacy (i.e., synthesizing and representing data). In the text, the data is presented in two tables. Have students consider how patterns might be easier to recognize if data were arranged differently. Ask students to individually consider ways to represent the data and record their ideas in their science notebooks. For example, a bar graph shows a comparison among categories. A pie chart is used to represent and compare parts of a whole. A bar graph that shows data in intervals is called a histogram. A line graph is used to display data that changes continuously over periods of time. Analyzing data to see how the climate has changed (and connecting that to CO_2 emissions) is supported by the Next Generation Science Standards and Engineering Practices (National Research Council, 2013) that call for students to analyze data systematically, looking for patterns, or testing whether data are consistent with an initial hypothesis.

Connecting to their previous analysis of line graphs, students might start by creating a line graph to represent the data. After creating at least two additional representations of the data, ask students to compare and contrast the graphs they created. In this activity, prompt students to discuss the following questions in small teams:

- Which type of graph shows changes between years most clearly when you compare 2012 and 2017 CO_2 emissions?
- Describe how the US share of global carbon emissions has changed between 2012 and 2017.
- Discuss which graph shows changes most clearly when you compare global shares between nations?

By analyzing the graphs presented by Fleischer in *A Hot Mess* and creating graphs of their own, students are analyzing data about climate change and learning how CO_2 emissions connect with climate change, specifically global warming and rising sea levels.

Sea-Level Rise

Exploration activities give students practice using physical models, collecting and analyzing data, as well as linking the scientific concepts of stability/change and cause/effect. On page 101, the book suggests the following activity as a means to aid in visualizing sea-level rise and the difference between ice melting in oceans compared to on land. This visualization can help students develop a deeper understanding of the impact of climate change on water, or the lack thereof.

We expand on the suggested activity in order to visually model and quantify the change in sea-level rise. We suggest students work together in teams of three or four, with each team provided with the following materials: a clear storage container, approximately 9" × 13" and at least 6" deep, rocks or sand that fit completely inside the container, water, ice cubes, clear transparencies, or a clear lid that fits the storage container, washable markers.

Part 1. To design their simulated environment, students begin by building an area of "land" in the storage container using the rocks and sand (see figure 2.3).

Figure 2.3. Part 1: Sea-Level Rise Investigation—Sea Ice. *Source*: Photo by K. Baldwin.

Students should then fill the container with water, being careful not to disturb the land areas, especially if sand is used. Following this, students will place twelve to fifteen ice cubes into the water area only. Next, before any of the ice melts, students tape a clear transparency or lid over the top of the container; this provides a topographic view of the model (see figure 2.4). Using a washable marker, students trace the shoreline onto the transparency/lid. This line is the "original" shoreline.

Next, students wait for the ice to melt and then retrace the shoreline with a second marker color (or symbols), making a key for their two, line colors/symbols, which enables them to track which line represents the original and which line represents the revised shoreline (see Figure 2.4). For example, black = original shoreline (Part 1); white = revised shoreline (Part 1). Note: the lines should be similar as there should be no rise in sea level in Part 1.

After students have traced the new shoreline, they will un-tape one side of the transparency and fold it out of the way or remove the plastic lid. With that out of the way, they can access their models to revise (i.e., conduct Part 2 of the experiment). One side of the transparency should remain taped to the side of the container so that it can be flipped back over again to trace shorelines during Part 2.

After students complete Part 1 of this activity, they will reflect on what they have learned about ice melt and connect what they've learned to chapter 4 in the text. This exploration develops a deeper understanding of how melting ice impacts land masses. In this case, there was little change when ice melt occurred in water; however, during Part 2 of this exploration, students will investigate the impact of ice melt on land, which more closely resembles the impact of ice melt in the polar regions—one of the impacts of climate change discussed by *A Hot Mess* in chapter 4.

Part 2. In this part of the experiment, students will place the ice cubes only on land, rather than in the water (see figure 2.5). Students can visualize the impact of sea-level rise by placing buildings close to the shoreline. In our

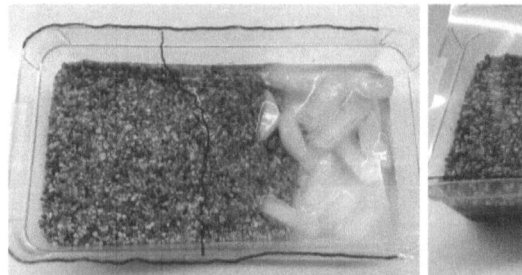 

Figure 2.4. Before/Original Shoreline (a) and After/Revised Shoreline (b). *Source:* Photos by K. Baldwin.

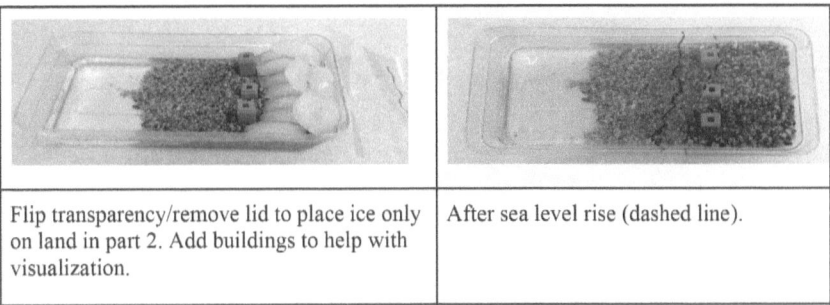

| Flip transparency/remove lid to place ice only on land in part 2. Add buildings to help with visualization. | After sea level rise (dashed line). |

Figure 2.5. Part 2: Sea-Level Rise Investigation—Ice on Land. *Source*: Photos by K. Baldwin.

case, we used measurement cubes to represent buildings. The students flip the transparency back over the top of the model so that they have a topographic view of the environment. They repeat the tracing process after the ice melting, using a different colored marker (or different symbols) and adding these additional colors to their key (e.g., black = original shoreline (Part 1), white = revised shoreline (Part 1), and black dashed line = revised shoreline (Part 2)).

Part 3. During Part 3 of the experiment, students analyze their data by removing the transparency/lid and examining their shorelines (see figure 2.6). The following questions could be considered:

- Describe what you notice about sea-level change when the ice was only in the water.
- Describe what you notice about sea-level change when the ice was only on land.

Have students refer to the figure on page 98 "Shrinkage of the Arctic Ice Pack" and respond to the following questions in their science notebooks:

- Based on what you learned from the previous sea-level activity, what impact does melting and shrinkage of Arctic ice have on global sea level?
- Based on the sea-level activity, what do you predict would be the impacts of melting ice in Antarctica on global sea level?

Students can review information on the following two websites selected from the "Further Information" section of the book (pp. 183–186).

- US National Climate Assessment—This website updates their climate report every four years. Impacts of climate change can be searched by region of the United States.

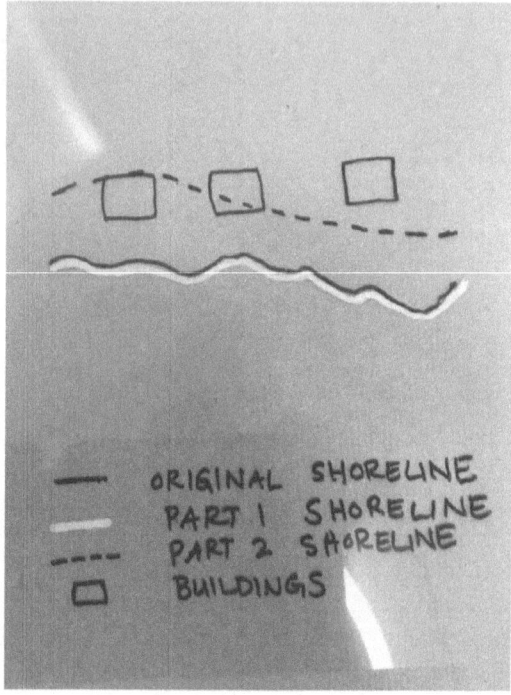

Figure 2.6. Transparency with Multiple Shorelines. *Source*: Photo by K. Baldwin. *Note*: Transparency was removed to show the three shorelines. As part of the data analysis, teams must develop an appropriate scale for their model (i.e., 1 inch = 10 feet).

- World Wildlife Fund—Conservation organization aiming to protect endangered species and their environments. Specifically, have students examine the WWF page "Six Ways Loss of Arctic Ice Affects Everyone."

Students can use the information from these sites to support their answer to the following question:

- Providing evidence from chapter 4 and the given websites, how do changes in global sea levels impact communities (e.g., chronic regular flooding, home displacement, etc.)?

To finish the "Sea-Level Rise" investigation, students examine the National Oceanic and Atmospheric Administration (NOAA, noaa.gov) Sea-Level Rise Viewer (as cited in Fleischer, 2021, p. 108). This simulation allows students to visualize the results (sea-level rise) of global ice melt. If your students are located in or near a coastal state, students can make local connections to the environment by examining local effects of sea-level rise. For interior states or to expand connections to global concerns about sea-level rise for all students, students can compare other countries or compare findings from multiple states' coastlines.

AFTER READING *A HOT MESS*

After completing both the *Engage* and *Explore* stages of the 5E model (Bybee, 2015), the *Explain* stage takes place. During this stage of inquiry, students can explain what they have learned throughout the first two stages (*Explore* and *Engage*) by creating an infographic that includes facts and details from their learning.

Infographic Explaining Climate Change

In science, students are often asked to present research and data in visual formats, such as posters or presentations, that include charts, tables, and facts. An infographic combines many of the elements of scientific literacy by including informative texts that explain scientific information, organizing facts and data to make connections, and using formatting and graphics effectively. This infographic project merges visual and verbal knowledge learned during the unit and connects with the *Explain* stage of the 5E Learning Cycle Model (Bybee, 2015).

To begin, the teacher shares expectations with students by providing a visual model. The sample infographic (see figure 2.7) shares information about climate change, as well as utilizes effective formatting for scientific writing. Culham (2014) coined the phrase "writing thieves" to describe the way that students borrow stylistically from published authors. Thus, students will use this model to guide their own work, "stealing" ideas for use in their own infographic.

To determine the essential characteristics of an infographic, Creative Educator (n.d.) suggests asking the following questions: *Which information, facts, and data are included? Which aren't? What colors and layout did they use? What graphs and graphics conveyed information and data to the viewer? What is the order, or flow, of information?* By examining the *ideas, organization, voice, word choice, sentence fluency,* and *presentation* of the model text, students apply knowledge of the six-traits of writing (Culham, 2014).

Of particular importance to this project, students analyze which key information about the environment is included; consider the colors and layout used by the author to draw readers' eyes to key data and facts; determine how graphs, graphics, and pictures are used to build knowledge about the

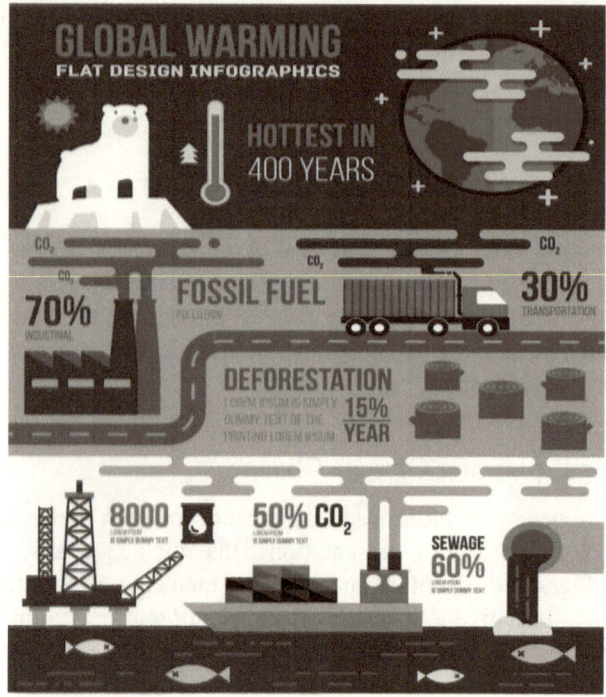

Figure 2.7. Infographic Model (template was downloaded from Creative Commons, Dooder, n.d.).

environment; and read the piece visually to analyze flow of ideas and information. In the sample, the infographic designer included percentages in large font throughout the infographic. This draws the reader's eye to these key facts. The colors are simple, focusing on red, white, brown, and blue, with some orange and gray. The brown might represent pollution. On the other hand, because the Earth is shown in shades of red and pink, it might be a clear representation of how polluted the Earth has become. In addition, the authors used a highway, as well as steam from a freight truck, ship, and power plant to connect parts of the graphic to one another. Additionally, the infographic is divided into three sections representing air, land, and sea. Students could select from one of the infographic templates available online or design their own.

After placing students in small teams, each team uses facts and data from *A Hot Mess* to create their own infographic based on a particular section of text (i.e., weather, fires, rising tides, mass extinctions, pollution, or climate migration). Since students have already read *A Hot Mess*, answered guided reading questions while reading, investigated specific components of climate change (i.e., "Exploring Sea-Level Rise"), and analyzed data (i.e., "Exploring

Data Literacy"), using information from the text itself will help students to explain what they have learned about climate science, as well as humanity's impact on the environment. As a reminder, the unit focuses on the following driving question: *What are the current, and possible future, impacts to the Earth's systems (biosphere, lithosphere, hydrosphere, and atmosphere) as a result of global climate change? Provide evidence from the text to support your answer.* Students build an answer to this driving question through the activities and through the guided reading questions in each chapter.

Teams create their infographic using a digital tool such as Canva or use one of the online templates available on freepik.com. If students don't have access to technology, they can create an infographic by hand using art supplies and 11 × 17-inch paper. Each team's infographic will be printed and posted in the classroom or hallway, and the class conducts a gallery walk to view other infographics. The teacher could provide post-it notes so that students can give feedback as they read and observe on another's projects. Each team will collect their post-it notes afterward so they can view the feedback provided by their classmates.

EXTENSION ACTIVITIES

The *Elaborate* and *Evaluate* stages take place after students *Explain* what they learned by reading *A Hot Mess*. According to Bybee (2015), the *Elaborate* stage is a chance to apply understanding of concepts developed during earlier stages of the 5Es and to develop a deeper understanding of concepts through additional activities. The *Evaluate* stage is a chance for the students and teacher to evaluate learning from the unit. In the following activity, we combine these stages with a Climate Showcase Poster presentation activity.

Climate Showcase Poster

After reading *A Hot Mess*, students should be *Evaluated* on their ability to explain, using an evidence-based claim, their answer to the driving question: *What are the current, and possible future, impacts to the Earth's systems (biosphere, lithosphere, hydrosphere, and atmosphere) as a result of global climate change? What is your evidence?* It is this question that guides their final project, a "Climate Showcase Poster," which is a gallery-style poster presentation modeled from the professional format that scientists present at conferences.

For this project, students will need to *Elaborate* on what they've learned by investigating additional questions and gathering additional research and data. Teachers assign students to small teams (ideally three to four students), and each team is responsible for one focus area of impact: deforestation, forest first, sea-level rise, or severe weather. Elaboration is the perfect opportunity for students to revisit unanswered questions (developed during reading). At this stage of the inquiry, the teacher could write all of the class' unanswered questions on the whiteboard. These questions could be used to guide teams as they investigate and research their final project. Using the "Selected Bibliography" (pp. 179–182) and "Further Information" section (pp. 183–186), students find and analyze at least two additional sources and data to support a claim about the impact their team has been assigned.

Finally, students present their findings in a Climate Showcase, a gallery-style poster presentation, where each team can present and then answer any questions that their peers may have about their topic.

CONCLUSION

Throughout reading *A Hot Mess*, students think critically about the impacts of global climate change such as wildfires, drought, rising sea levels, and deforestation. The approaches and activities presented in this chapter guide students through a scientific inquiry using the 5E Learning Cycle (Bybee, 2015) and the young adult text *A Hot Mess*. When finished, students should be able to describe the effects of global climate change and the impact it has on communities. The reading activities integrate literacy and science by asking students to build on prior knowledge, engage and analyze visual representations of data, connect textual evidence from the book, explore impacts of global climate change, and, ultimately, build their environmental literacy. Having students build their content knowledge about the environment will directly impact how those students interact with the environment and make informed decisions, as it relates to the environment and throughout their lifetime.

REFERENCES

Bybee, R. (2015). *The BSCS 5E instructional model: Creating teachable moments.* NSTA-Press.
Culham, R. (2014). *The writing thief.* Stenhouse.
Dooder. (n.d.). Global warming infographic template. *Free Vector.* https://www.freepik.com/vectors/infography.
Fleischer, J. (2021). *A hot mess: How the climate crisis is changing our world.* Zest Books.
Fulton, L., & Campbell, B. (2014). *Science notebooks: Writing about inquiry.* Heinemann.

National Academies of Sciences, Engineering, and Medicine. (1996). *National science education standards*. The National Academies Press. https://doi.org/10.17226/4962.

National Research Council. (2013). *Next generation science standards: For states, by states*. The National Academies Press. https://doi.org/10.17226/18290.

National Science Teaching Association. (2016). *Teaching science in the context of societal and personal issues*. NSTA Position Statement. https://www.nsta.org/nstas-official-positions/teaching-science-context-societal-and-personal-issues.

National Science Teaching Association. (n.d.). *Science and engineering practices*. National Science Teaching Association. https://ngss.nsta.org/Practices.aspx?id=4.

Zucker, A. (2021). Teaching scientific literacy. *The Science Teacher, 88*(4), 8–9.

Zygouris-Coe, V. I. (2015). *Teaching discipline specific literacies in grades 6–12: Preparing students for college, career, and workforce demands*. Taylor & Francis.

Chapter 3

Countering "Plant Apathy"

Using Kenneth Oppel's Bloom as a Motivating Tool for Teaching Plant Science to Students

Katharine Covino and Erin Rehrig

Wandersee and Schussler (1999) first coined the term "plant blindness" over twenty ago to explain an overall civic and educational lack of awareness and appreciation for plants and their importance in the environment. This apathy often begins in early elementary school and persists into college. We are biologically drawn to facial features and often find animals more interesting. For example, if you show students a video of an orangutan in the jungles of Borneo, they might focus on the animal and overlook the lush, green biodiversity of the surrounding forest. Unfortunately, overall content knowledge about plant science is also deficient and a general lack of motivation toward botanical subjects impedes students' abilities to learn new material about plants (Staag et al., 2009). Because plants don't move from place to place like animals do, often students find them irrelevant or boring (Sundberg, 2008). This creates misconceptions that plants are physiologically inactive and can make teaching plant biology a daunting task.

How can we help adolescent students become more interested, knowledgeable, and passionate about learning about plants? One way is through the use of young adult (YA) literature. Although English and History teachers have long seen YA literature as a way of connecting to and supporting content-area teaching and learning, other subject areas, including science, are only beginning to embrace the use of YA texts in their subject areas. Now is the time for science teachers to find value in contemporary books written for teens (Roberts et al., 2012). Many YA novels offer students an opportunity to engage in reading while learning about scientific concepts;

Bloom (Oppel, 2020) is one such novel that offers students a chance to learn about plants and their importance in the ecosystem.

Kenneth Oppel's book *Bloom*, intricately plotted and meticulously researched, can motivate students to learn critical plant science topics such as photosynthesis, plant anatomy and growth, and the biology of invasive species. *Bloom* encourages students to envision plants as active participants in the ecosystem and not as non-living or uninteresting organisms. With a gripping plot and apocalyptic setting about invasive alien plants, *Bloom* interjects accurate botanical topics throughout. While students are engrossed in this thrilling novel, science teachers can harness interest and excitement to engage students in plant biology. The instructional approaches suggested in this chapter meet many science standards that introduce the concepts of plant biology, including photosynthesis, seed dispersal, and growth as well as the scientific method using inquiry-based methods.

BLOOM BY KENNETH OPPEL

Bloom opens with a view of a world remarkably similar to our own. Three high school students struggle to navigate school, family, and friendships. Petra, popular but fragile with her crippling allergy to water, feels that her many friends don't truly understand her. Anaya, smart but diffident, lacks confidence because of her appearance. Seth, the new kid in school, desperately wants a family to call his own after years in the foster care system. While they attend the same school, their lives seem disconnected—until the rains begin.

An intense, soaking rain storm leaves in its wake seeds that quickly and aggressively begin to terraform the planet. The fast-growing plants thrive in their new home and quickly begin to choke out all native crops and vegetation. What's more—they seem impossible to kill. The plants then begin devouring animals, including humans. As the invasive, sentient plants take over, Petra, Anaya, and Seth realize that they are changing, too. While most people suffer from contact with the plants, the three adolescents are seemingly immune. Bound together by their strange, emerging gifts, they seek to use their connection with the alien plants to save mankind. But will their friendship and their strange, new mutations be enough to stem the tide of the invasion?

PREPARING STUDENTS TO READ *BLOOM*

What's in a Word?

Prior to reading Oppel's YA novel, teachers can help their students closely and critically engage with the single word "bloom" that serves as the title of the text. Encouraging students to think about the multifaceted nature of the word "bloom" in light of both their lives and experiences and also within the context of plant biology, teachers can guide their class in a lively, engaged conversation exploring the meaning of the word as a verb, noun, and symbol. To facilitate and capture this conversation, students can use a graphic organizer like the one suggested in figure 3.1.

Judging a Book by Its Cover

Following the exploration of the word *bloom*, teachers can guide students in an examination and analysis of the cover of the novel. Provide students with a copy of the cover and ask them to draw a vertical line down the middle and a horizontal line across the middle which will create four separate squares. Ask students to consider what they see in each square: *What are the colors used? What shapes are there? What images or words are present? What draws the most attention?* Have students annotate their responses directly on the cover. After students have responded for each square, ask them to share their responses with a small team. Students should then look for patterns across the responses. Once patterns are identified, students can explain what the pattern does for them as readers and consider what clues the cover might be giving them about the content of the book. This can be done in small teams or as a whole class.

Thinking Like a Botanist

There are key terms that are necessary to learn in order for students to understand the botanical concepts explored in the book. Understanding these terms beforehand will help students become better prepared to compare the plants'

What does "bloom" mean to you? Draw an image below.	Share your drawing with a classmate and reflect on the comparison.	Look up and at least 2 definitions of the word bloom below.	Indicate the part of speech for each definition.
	We both drew flowers with petals.	1. The flowering state 2. Reaching your potential	1. Noun 2. Verb

Figure 3.1. Sample Graphic Organizer (completed by authors).

features presented in *Bloom* with those found in our world. In addition to positioning students to compare traits, learning the academic language of plant biologists can also help students understand both features and functions of plants—specifically, how plants acquire nutrients and resources needed for survival, how they grow, and how they reproduce. Table 3.1 offers a complete list of botanical vocabulary found in the book. Many of the terms can be defined using context clues from the book. To learn the vocabulary terms, students can visit the page where each word appears, review the context, and record their findings on a flashcard with an illustration. Students can learn and internalize these words by working with a partner using their illustrated flashcards to play a game of memory at the start of the next class as a warm-up. To play memory, students simply place all flashcards face down on a table and each takes a turn flipping over two at a time. The goal is for the student to match the flashcards. Once they are matched, the flashcards are removed and placed to the side. After the game, teachers could use students' cards to create a word wall.

Creating a Lab Notebook

Several suggested activities in the during-reading section of this chapter require students to create and keep a lab notebook. As such, it is important for teachers to explain the expectations for creating this notebook. Notebooks are critical scientific tools that students can use to journal their observations, work through calculations, create illustrations, and write reflections. For example, students will use these for drawing and making notes while dissecting and observing Earth plant features to compare the ways that the fictional plants in the novel are both similar and different from the ones they are observing in the lab as they read. By emulating what real scientists do in the field, students' use of lab notebooks can be a powerful scientific pedagogical tool, helping them articulate their understanding of a topic and take ownership of their work.

WHILE READING *BLOOM*

As students progress through Oppel's text, teachers can guide their thinking around the concepts of anatomy, reproduction, growth, and seed dispersal. The scientific activities suggested in this section align with concepts explored in the text: dissections, scientific inquiries, and time-lapse photography align to these concepts in the text. In an effort to hold students' interest and motivate them to learn more about plant biology, the

Table 3.1. Scientific and Botanical Vocabulary Found in Bloom

Vocabulary Word	Definition	Pg. #
Bloom	To flower, or open up	9
Unruly	Uncontrollable, stubborn	10
Sheath	Protective covering surrounding living tissue	22
Rhizomes	Underground shoots a plant uses to give rise to new plants through vegetative propagation	44
Invasive Species	Species of plant found in a non-native environment that grows prolifically and destroys natural habitats	45
Allomones	Chemicals released into the soil by plants to deter other species of plants from growing nearby	46
Photosynthesis	Process by which plants utilize the energy from sunlight to convert inorganic carbon dioxide (CO_2) and water (H_2O) into sugars and oxygen (O_2) needed for growth	48
Wavelength	Distance between light wave peaks in the electromagnetic spectrum which affects the amount of energy and color	48
Germinate	When seeds become active, leaving a dormant stage and sprout into a small seedling	52
Herbicide	Chemical toxic to plants used to curb or prohibit growth	53
Undifferentiated	Tissues or cells that have the ability to become any type of tissue or organ (stem cells)	72
Tendril	A modified type of leaf on a vine that curls around an object to allow the vine to climb upward	152, 251
Enzyme	A protein used in digestion of food	156
Alien	Foreign or not native to a particular region or planet	157
Cryptogenic	Of unknown origin	158
DNA	Deoxyribonucleic acid, or the genetic material found in an organism's cells	158
Amino Acids	Small organic molecules that are the building blocks of proteins necessary for life	158
Genome	The complete sequence of DNA in an individual's cells that differs from others	169
Aquagenic Urticaria	A rare genetic mutation that causes an allergic reaction when water contacts the skin	171
Methane	CH_4, a highly flammable gas with a rotten-egg smell often produced in bogs and swamps as the product of decay	259
Fungal Ecosystem	A specialized network of fungi and other organisms that live mostly underground and help plants acquire nutrients	264
Bacterium	A small single-celled organism	264

during-reading activities offered in this section are designed to address the overarching discussion question, *How are the structures and functions of Earth plants similar to and different from the alien invasive species in the book?*

How Pollen Gets Dispersed

One point of contrast between Earth plants and the fictional invasive plants in *Bloom* has to do with how they emit pollen. Early in *Bloom*, the alien plants wreak havoc on the towns peoples' allergies due to the copious amount of pollen they spew (p. 67). In this activity, teachers can guide students as they dissect various pinecones (male pine pollen cones) or flowers such as Easter lilies with large sweet, sticky-smelling, and pollen-producing anthers. Following the dissection, teachers can support students as they make a wet mount of pine or lily pollen on a glass slide to view under a microscope. Students draw what they see and use multiple fields of view to calculate how many pollen granules are in a single water droplet. This connects back to Anaya's observation in the book about the alien plant's ability to disperse its pollen; "The long black stem arched and swelled. The flower trembled and then snapped forward. With a loud pop that made Anaya jolt, a thousand tiny grains of pollen exploded through the air" (p. 68). Students will discover that there are millions, not thousands, of pollen grains in a single flower. This deepens their understanding of both of Earth plants' reproductive processes, while also helping them envision the inescapability of the pollen emitted by the alien plants in the text.

How Berries Look and Function

Another point of comparison between Earth plants and the invasive alien species is the way each produces berries, which function to support life and further the propagation of the plants. In *Bloom*, pit vines plants produce juicy fruits for attracting animals that are then captured and digested. A third of the way through the novel, as Petra and Seth walk together, they are distracted by a dark, deep purple spring of berries (p. 105) which look delicious. To further explore this topic, students dissect their own bright berries or fruits. Teachers can easily find blackberries, grapes, or strawberries for students to dissect. As students dissect each fruit, they can return to their lab notebooks to capture details of the plants' features. In addition to exploring the berries' appearance, teachers can help students to explore the different ways they function. On Earth, unripe blackberries, grapes, and strawberries are green and hard, however, during the ripening process; they fill with sugars and become red (or purple). This signals to birds and other animals that they are sweet, nutritious, and ready to eat. Most animals, especially birds and humans, have excellent vision for seeking out red berries. Birds digest the berries and then defecate the seeds miles away, which helps disperse the seeds away from their parent plants. Investigating, dissecting, drawing, and discussing these berries can help students understand that the plants in *Bloom* use the same strategies that Earth plants use—to attract birds and other animals to their ripe fruits, albeit for different reasons. Earth plants engage in this process to further seed dispersal, whereas the carnivorous alien plants in *Bloom* use berries to lure in their prey. For example on pages 102–103 a small bird is engulfed by the pit vine plant after plucking off a bright, juicy berry. This activity connects to and reflects the text by helping students understand and visualize the importance of plants producing brightly colored fruit to attract animals.

Becoming Pant Biologists through Scientific Investigations

Within the opening pages of the book, Oppel introduces a supporting character, Anaya's father, who is the primary source of plant professional and academic knowledge in the book. His character is a botanist with the Ministry of Agriculture who works at the island's experimental farm (p. 10). He provides intelligence and insight and leads the struggle to understand, and ultimately battle, the mysterious invasive plant species. For example, his ability to simply, yet effectively, explain the concept of allomones to Anaya; "Yeah, it's like chemical weapons. The garlic mustard wages warfare in the soil. It releases chemicals from its roots to keep anything else from growing. Basically, it poisons the soil for other plants" (p. 47).

Scientific investigations in the classroom also provide a model for student readers because they mimic the work of real-life plant biologists and highlight the central importance of the scientific process through experimentation and investigation. Mirroring the work of the experts in the text, teachers can guide students in various inquiry projects that allow students to explore first-hand some of the plant biology concepts discussed in *Bloom*. The activities below are designed to help students engage with and better understand the concepts of photosynthesis and seed dispersal. These activities support teachers who are interested in engaging in inquiry-based scientific experimentation with their students.

Photosynthesis

One of the startling traits of the alien plants depicted in *Bloom* is their unusual color. Early in the novel, Anaya's father mentions that the cryptogenic plants are truly black, "which means they can photosynthesize at any wavelength of light" (p. 48). To deepen students' understanding of the process of photosynthesis and to enable them to hypothesize which color light in the visible spectrum (red, orange, yellow, green, blue, indigo, violet, and white) causes the fastest rate of photosynthesis in spinach, students can carry out a simple but telling experiment. The photosynthesis lab we describe is inexpensive and simple for students and teachers to set up. Furthermore, the activity can be performed in a typical fifty-minute science time block within a school's schedule. To start, students use paper hole punchers to make ten spinach leaf disks which they then place in a 10-mL syringe (no needle) with the plunger removed. Students then replace the plunger, draw up 5–8 mL of water, tip the syringe right-side up, and press the plunger to remove all the air. With their fingers covering the tip, have students slowly pull back on the plunger for two to three minutes. This creates a vacuum and removes all of the gasses trapped inside the layers of the plant leaf disks. After the air is removed, the disks will sink. Have students empty the disks into a beaker containing 50 mL of a 1 percent baking soda solution, which serves as the main source of carbon dioxide. To prevent light exposure (and the start of photosynthesis), the beakers should be covered with foil until the experiment is ready to begin. Students then place exposed beakers with disks under a lamp for thirty minutes. Control teams should use white light, while other student teams experiment with red, blue, green, or no light at all. The disks should begin to float due to oxygen gas that is created when the plants use the energy from the light to convert carbon dioxide and water into sugar and oxygen in the process of photosynthesis. Here, ask students to record the number of disks floating after ten, twenty, and thirty minutes and calculate a final percent. From these data, students can determine which color of light allows spinach plants to best undergo photosynthesis. Furthermore, they can now critically think about why the black, cryptogenic plants in the novel have a greater advantage over Earth plants in their ability to utilize *all* the colors of the visible electromagnetic spectrum.

Seed Dispersal

The plants Oppel depicts in *Bloom* also stand out for their ability to disperse their seeds in a hyper-aggressive way. For example, the cryptogenic water lily plants eject seeds by spitting them out like acidic bullets from a large flower head. One of the main characters, Petra, uses this same feature to destroy the plants in "Rambo-esque" style,

> She grabbed another lily head from behind. She felt it trying to turn on her, but she held tight, and spun around, spraying its seeds at the other plants. Severed (flower) heads hit the water. Beside her, Seth surfaced, gasping, and stared at her in amazement as she machine-gunned more lilies. (p. 281)

To help students understand the purpose and process for seed dispersal (in both real and fictitious plants), teachers can guide them in a hands-on, investigative activity, described by Wampler and Dobson (2010), that asks them to gather, examine, and experiment with seeds. To begin this hands-on activity, students gather local seeds (e.g., maple samaras, a.k.a. helicopters, which are largely geographically ubiquitous) and measure their weight and total "wing" area. Next, students drop the samaras from an elevated location and measure dispersal distance from different heights then graph area and weight against height. The goal of graphing is to help students understand the role that seed structure and function has in successful dispersal far from an original drop site or parent plant. Once the experiment is over and all drops have been graphed, students complete a final lab report that includes an introduction, hypothesis, methods, results, and discussion of seed dispersal. This report would demonstrate the students' growing knowledge of seed dispersal patterns of Earth plants and could also be used as a comparison to the ways the alien plants disperse their seeds. This also reflects how scientists (like Anaya's father and Dr. Stephanie Weber, another character in the book) communicate their findings to the larger community.

Fast Growth

The alien plants in *Bloom* grow very quickly. There are numerous instances throughout the novel documenting the staggering rate at which the alien plants grow and transform (see table 3.2). The plants grow incredibly fast in a number of conditions on the small Canadian island and around the world (see table 3.2). This ability is one of the features that makes them such a daunting foe.

To help students visualize and interact with fast-growing Earth plants—to actually see them grow—teachers can obtain Wisconsin Fast Plants or other available seeds that grow quickly. With assistance from the class, the teacher can set up a time-lapse camera or iPhone programmed to take a photo every ten to fifteen minutes. These real-time photos will make clear the process and speed from which the plant grows: from seed to senescence. Once the

Table 3.2. Textual Examples of Plant Growth and Transformation in Bloom

Page #	Quotation from Bloom Discussing the Fast Growth of Plants
35	"It was tall and black and spiky, and seemed to have shot up overnight as the rain hammered on the roof."
52	"Nothing normal grows that fast."
74	"She didn't know much about plants, but you weren't supposed to be able to *see* them growing."
153	"That just grew overnight?"
224	"This is already more dangerous than we thought. I've never seen the vines growing so fast."

activity is complete, students create a movie to see how fast plants can grow. These artifacts can inform a whole-class discussion regarding how people's perception of time may be different from plants' perception of time. To deepen and reinforce students understanding of these concepts and the links to *Bloom*, teachers might ask some of the following higher-order guiding questions:

- In what ways do plants grow on the same timeline as humans?
- In what ways might this timeline be the same for all plants? How might it be different? For example, how do fast plants grow compared to a Sequoia tree?

Capturing the growth of Earth plants in real time and comparing the speed to that of the alien species in the novel will not only help students understand the concepts of plant growth but lend a greater appreciation of what the characters in the text are up against.

AFTER READING *BLOOM*

Building from Expertise—Synthesizing Data to Answer New Questions

Characters in *Bloom* learn about alien plants and themselves. Drawing from inquiry and analysis, Petra, Anaya, and Seth build their knowledge through experience until they are able to understand and theorize more about the invasive species. Similarly, through the process of reading the book, students have had an opportunity to explore and discover more about plant biology. By keeping careful records of vocabulary, illustrations, timelines, data, calculations, and reflections in their lab notebooks as they read, students should envision the way the alien plants look and function. From their active, hands-on engagement with the various activities while reading, students identified ways plants, both real and fictional, function and grow. In short, by adopting the language and practice of plant biologists, students have gained invaluable experience and expertise. They are ready to apply their growing understanding to a new type of task—theorizing from data.

In the novel, it becomes clear to the central three characters that alien plants are both sentient and intelligent. Early in the story, Anaya feels "keenly that the grass had an intelligence" (p. 68). Much later, the devious and deadly intelligence becomes less subtle and far more terrifying. As Petra realizes, "Those vines. They were psychotic. They could bring down a helicopter. They could strangle all of them" (p. 239). The characters use their observations to construct theories about the alien plants. In the following activities, students can follow the same process. Well-versed in how plants look and how they function, students are now primed to synthesize from their various sources of knowledge and begin to discuss their views concerning plant intelligence.

Are Plants Intelligent? Podcast Project

The denouement of the novel reveals the extent to which the alien plants are both intelligent and sentient. They understand their environment. They have strategies to take on their opponents and are able to communicate (pp. 274–275). After reading the novel, teachers can guide students as they move from analyzing plant features and functions to asking larger questions regarding plant intelligence by having them create small-team podcasts. Specifically, this suggested after-reading activity asks students to build on what they have learned about Earth plants to explore the ways such plants adapt to their environments. As highlighted in table 3.3, many Earth plants have unique abilities and outward characteristics that often make them intelligent and sentient—in some cases more like slow-moving animals (Schultz, 2002). In fact, there is evidence that plants can respond to both the sound of caterpillars feeding on them (Appel & Cocroft, 2014) and the smell of airborne chemicals from nearby insect-attacked plants (Heil, 2014). As a way for students to further develop their knowledge about plants and connect this knowledge to the plants in *Bloom,* students can create podcasts. Podcasts are quickly becoming popular sources of digital media for schools. Fortunately, due to the creation of free online audio capture and editing tools, as well as built-in features of most cell phones and laptops, the financial costs and technological requirements of creating a podcast can be nominal. For example, free downloadable software such as Spotify's Anchor, Apple's Garage Band, or Google's Meet App, just to name a few, offer students the ability to record directly from their phones or computers.

For this after-reading activity, place students in teams to create a ten-minute podcast that addresses the following questions:

- What does it mean to be an intelligent or sentient being?
- How do plants "see" light and why is this important?
- What other senses (smell, taste, hearing) are plants capable of? What functions do these senses serve?
- What evidence is there that plants can communicate with other plants?
- What evidence is there that plants can communicate with insects or other animals?
- How do communication signals above ground (through leaves) vs. below ground (through roots) differ?

Table 3.3. Comparison of Cryptogenic Plants vs. Earth Plants

Page #s	Cryptogenic Plant Characteristics	Earth Plants with Similar Abilities (common names only)
22, 48, 153	Rapid growth (all three plants)	Bamboo, Phragmites, Mile-a-Minute Vine, Kudzu, Grapevine
40, 115, 120, 241	Strong. Requires chainsaw or knife to cut through (black grass, viney pit plant)	Ash, Hickory, Ironwood Tree (Australia)
47	Produces allomones (pit plants)	Spotted Knapweed, Black Walnut
49	Have dark, almost black color (all three plants)	Black Mondo Grass, Bat Orchid Flower
51, 71	Seeds germinate in water (all three plants)	Tomato, Lettuce
55, 57	Burn like oil and produce thick, irritating smoke (black grass)	Acacia, Eucalyptus
58	Resistant to herbicides (black grass)	Ragwort, Wild radish
68	Produce copious amounts of pollen and has pollen flowers that "pop" (black grass)	Birch, Pine
73	Produce irritating acid or chemical that ***instantly*** burns or blisters flesh (water lily)	Stinging Nettle, Giant Hogweed, Manchineel
103, 105	Produces sweet fruits or nectar and uses bright colors to attract animals into a pit (vines of pit plants)	Pitcher Plants
103, 112, 139, 156	Traps animals in a "sac" and uses digestive acids and corrosive enzymes to consume animal material (pit plants)	Pitcher Plants, Venus Fly Traps, Sundews
106, 121, 151	Produces a perfume or chemical that is able to induce sleep and/or dilate the pupils	Poppy, Deadly Nightshade, Valerian
127	Plants communicate with each other using volatile chemical signals (black grass and pit plants)	Alder, Lima Bean, Wild Tobacco
139	Vines/plants that are difficult to pull off or grow on top of other plants to choke them out (vines of pit plants)	Virginia Creeper, Ivy, Bittersweet, Kudzu, Strangler Fig
224	Growing in a particular pattern near or away from a stimulus in an "intelligent" manner (vines of pit plants)	Dodder
248	Dispersing seeds from a seed pod violently almost as if "spitting" them out (water lilies)	Touch me not, Squirting Cucumber
81, 241, 248	Having highly acidic seeds or roots that acidify the water (water lily)	Avocado, Corn, Rice
259	Produces or "exhales" methane (CH_4) into the atmosphere (water lily)	Rice, Marsh plants

- Can plants feel pain?
- Based on the evidence you provide, do you consider Earth plants to be intelligent? Why or why not? What about the plants in *Bloom*?

Each team can be assigned three different questions, with all teams addressing the first and the last question. In order to successfully complete their podcast, student teams can conduct research together using library resources and science article digests. One student in each team should serve as the host who facilitates questions, creates a simple script, and summarizes information, while the other students in the team serve as science "experts," citing evidence from their team's research. Once the podcast is completed, have students save it as .wav or .mp4 file and post them to the course website or send them to the teacher directly. The entire class can then listen to all podcasts and as a whole class discuss the information shared in each. Through this activity, students will provide critical and valuable feedback to their peers while also promoting a researched-based scientific discourse on plant biology.

"Terraformed" Hallway

The battle between the humans and the sentient plants is truly one of the most compelling and exciting sections of the novel. While much of the novel has shown the intelligent, alien plants with the upper hand, the tide turns when Anaya's father, the botanist with the Ministry of Agriculture, determines that a certain type of soil contains a bacterium that can kill the plants (p. 218). As his daughter and her friends make their way to the island to rescue him and make their stand against the alien species, a terrifying epic confrontation erupts (p. 224). To capture the drama and importance of this battle between the humans and plants, teachers can guide students as they use their creativity to create colorful, life-size cutouts (with their anatomical features labeled using the vocabulary from the pre-and during-reading activities). Applying an understanding of the plants—a deep awareness of the features, functions, and deadly intelligence of the plants—students can choose to design black grass, viney pit plants, or water lilies. Plant drawings can be displayed in one hallway of the school that is also decorated with artificial vines or streamers, warning signs, and water lily bogs. For fun, students can also spray "prank rotten-egg smell" on the cutouts to simulate the methane gas and hydrogen sulfide produced by the roots of the water lilies. In essence, students are recreating the book's setting in a hallway or classroom of the school, mimicking how the cryptogenic plants are terraforming the planet.

EXTENSION ACTIVITIES

Exploring Your Own Backyard

Many of our suggested tasks focus on the interactions between the characters and plants in the novel. The extension activities, however, ask students to make clear connections between what they have learned from *Bloom* and apply it to a close and careful study of the local environment. An interesting lens to help students make the jump from the confines of the book to the real world around them is by investigating the notion of *invasive species*. In *Bloom*, the cryptogens are the ultimate examples of invasive plant species. They are successful because when they land on Earth, they encounter no natural predators or competition. In fact, the alien plants are seemingly immune from the allomones secreted by the native-born species,

> Allomones? Sometimes Dad forgot that not everyone knew all the special plant words he did.
> Yeah, it's like chemical weapons. The garlic mustard wages warfare in the soil. It releases chemicals from its roots to keep anything else from growing. Basically, it poisons the soil for other plants.
> But not the black grass, Anaya said.
> No. Seemingly, black stuff's immune. It just out-bullied the bully. (pp. 46–47)

Leaping from the book to the real world, global issues of invasive species can often be seen right in our own backyards. Here, teachers can introduce the topic of invasive species that exist in North America and around the world. Invasive species thrive in new environments due to their ability to adapt to the new surroundings, sequester resources quickly, evade predators and pests, and even chemically alter the new environment, making it unsuitable for native plants. For example, showing before/after photos of natural areas that have been taken over by invasive or exotic plants may be a highly effective and visual strategy to help students understand their ecological impact. Specific case studies include cattails and purple loosestrife taking over the Florida Everglades and Eurasian water milfoil filling up lakes of the Pacific Northwest. An additional example is spotted knapweed (native to Asia), which grows prolifically in the Western United States. Like the black shoots in *Bloom*, knapweed produces and secretes allomones, chemicals which spread into the soil and kill the roots of neighboring plants. Even after many years, no plants or grasses are able to grow in the altered soil. It is a bane to farmers just like the black grass is in the book. A logical and fun extension activity that can follow the book is a field trip or nature walk around the school or local area in search of invasive plants. Unfortunately, many invasive plants are so ubiquitous that it will not be difficult to find them. Students can use photos, field guides, or internet resources to help with plant identification. For example, in *Bloom*, the townspeople use an app called Plant ID to try to identify the plants using crowd-sourcing (p. 43). For this extension activity, students can use Plant ID or a similar app called Seek by iNaturalist to help them identify local invasive species. This app allows users to take a photo of a plant with their phone and uses crowd-sourced data to identify it, often to the *Genus* level.

Once students find various examples of local invasive plants, they can take photos, make observations, and take field notes. Teachers can then assign each student or team an individual invasive plant species found during the walk. Students can then create a "WARNING" poster using poster paper and art media, Google Slides, or Canva. Students should research their invasive plant using vetted websites such as the Audabon Society, The United States Geological Survey, or the North American Invasive Species Network. On the poster, students should describe the plant's origin, life cycle, problems, and of critical importance, and potential treatments. Ask students to consider what is being done to eradicate that specific invasive species. For example, at the end of *Bloom*, scientists use pathogenic soil bacteria as an herbicide to destroy the alien plants. Interestingly, this technique is an example of an integrated pest management strategy that is often used for eradicating invasive species across the globe. Posters can be displayed in the classroom or hallway and can be shared at an open house night for parents.

CONCLUSION

Bloom is chock-full of science content about plants. Reading this text with students can empower biology teachers to offer their classes a unique opportunity: merging the motivating activity of reading YA with the engaging hands-on inquiry of science experiments. This text, and the accompanying unit-long activities and resources suggested in this chapter, solve a problem frequently faced by biology teachers. Too often instructors struggle to build exciting lessons about plants and botanical concepts due to ingrained and long-standing preconceived notions that plants are boring and uninteresting. *Bloom* challenges these staid ideas. Combining a thrilling plot, diverse and dynamic characters, together with a carefully researched exploration of the ways plants exist, develop, and evolve, the book helps students see plants in a whole light. No longer relegated to the background or the sidelines, the plants in this book upset preconceived notions. In this apocalyptic story, the invasive alien plants are intelligent, sentient, powerful, and villainous. With their collaborative exploration of the text, students will lose their "plant blindness" and gain a deep appreciation of the fascinating world of plants around them.

REFERENCES

Appel, H. M., & Cocroft, R. B. (2014). Plants respond to leaf vibrations caused by insect herbivore chewing. *Oecologia, 175*(4), 1257–1266.

Heil, M. (2014). Herbivore-induced plant volatiles: Targets, perception and unanswered questions. *The New Phytologist, 204*, 297–306.

Oppel, K. (2020). *Bloom*. Alfred A. Knopf.

Roberts, M., Dashner, J., Bodeen, S. A., Pfeffer, S. B., Crowe, C., & Albom, M. (2012). Teaching young adult literature: Why should we have all the fun? Encouraging colleagues to read YA novels across the curriculum. *The English Journal, 102*(1), 92–95.

Schultz, J. C. (2002). Biochemical ecology: How plants fight dirty. *Nature, 416*(6878), 267.

Stagg, P., Wahlberg, W., Laczik, A., & Huddleston, P. (2009). *The uptake of plant sciences in the UK: A research project for the Gatsby Charitable Foundation.* https://www.gatsby.org.uk/uploads/plant-science/reports/pdf/cei-uptake-of-plant-science-in-uk-feb-09.pdf.

Sundberg, M. D. (2008). Plants really are B.O.R.I.N.G. *Plant Science Bulletin, 54*(2), 50–55.

Wampler, L., & Dobson, C. (2010). Helicopter seeds and hypotheses . . . That's funny! In *Tried and true: Time-tested activities for middle school.* National Science Teachers Association.

Wandersee, J. H., & Schussler, E. E. (1999). Preventing plant blindness. *The American Biology Teacher, 61*(2), 82–86.

Chapter 4

Making Botany Magical

Teaching about Plants with This Poison Heart

Julie C. Baker, Shawn E. Krosnick, and Kelly Moore

As our planet experiences the effects of climate change, human population growth, and biodiversity loss, K-12 educators are tasked with finding new ways to develop science literacy and inspire students to take an active interest in the natural world. Career opportunities for graduates with training in environmental science, conservation biology, and bioengineering are increasing rapidly and will only continue as demand for new approaches and solutions grows. This need is affirmed in *A Framework for K-12 Science Education* (National Research Council, 2012), where the authors note the percentage of students motivated to pursue careers in science and engineering is too low for the nation's needs. The *Framework* also emphasized the importance of students understanding the role humans play in harming—and helping—the Earth. Specifically, a strong emphasis is placed on concepts such as ecosystems, biological evolution, biodiversity, and the impact of humans on the environment. The urgency in preparing students to meet global challenges has refocused science education toward the natural world and the importance of understanding these complex issues.

    Yet with increased emphasis on biodiversity and ecosystems, educators still struggle to successfully address one of the most fundamental components of ecosystems on Earth: plants. All of life depends on the oxygen that photosynthesizing plants produce. Plants are primary producers, meaning they support entire food webs by turning sunlight into usable energy in the form of carbohydrates. As plants grow, they remove carbon dioxide from the atmosphere and lock it in solid form, thus helping to reduce greenhouse gasses. Bar-On et al. (2018) estimated the biomass of different groups of organisms on Earth in terrestrial and marine habitats. Plants account for 81 percent of all biomass (450 gigatons) on Earth, while animals account for just 0.3 percent (2 gigatons). Even with plants occupying so much of the available space on Earth, they still fade into the background to humans, who tend to over-value vertebrates relative to other forms of life (Knapp, 2019). Given the stakes, finding meaningful ways to engage students in botany is essential to developing the next generation of citizens that will develop solutions to these challenges.

The question remains: *How do science educators effectively engage students in learning about plants and the natural world more broadly?* In the classroom, science literacy may be undermined by the use of traditional academic content that is dry and dense. As an alternative, Boswell and Seegmiller (2016) suggested using fiction as a means for students to form personal connections to content and relate that back to required material. *This Poison Heart* by Baylon (2021) is an engaging young adult novel that draws readers into a magical world of plants and the people tasked with protecting them. References to botanical content throughout the text provide learning opportunities and serve as starting points for activities and discussions on diverse aspects of science literacy. Through the pairing of a high-interest, captivating story about a young woman and her family history and science content incorporated throughout, this novel strikes a perfect balance of reading for both enjoyment and learning.

THIS POISON HEART BY KALYNN BAYRON

Briseis, also known as Bri, has a mysterious secret that she must keep from her friends and the public—a secret even she doesn't fully understand. Her secret is that she possesses a supernatural gift that enables her to make plants grow in response to her touch, her thoughts, and even her emotions. Her parents know about her gift, but they also agree she must be very careful with her unique power over plants. She tries to enjoy a typical teenage life, but the plot thickens when her aunt dies and wills Bri and her two moms a mysterious and deadly estate in upstate New York. As Bri explores the grounds that she inherited, she discovers she may not be the only person in her biological family with something to hide. For years, her family has grown a garden of secrets—a magical garden full of dark forces. Briseis' plant magic manifests like never before. She begins to uncover a dangerous world of potions and poisons while at the same time she is faced with keeping herself and her family safe.

PREPARING STUDENTS TO READ *THIS POISON HEART*

Botanical Terms and a Guided Dissection

This Poison Heart references botanical terms on nearly every page of the text. Plants are central to the story and an essential part of Bri's character development. To help students better appreciate the botanical references in *This Poison Heart*, teachers may wish to review basic plant terms with students (see table 4.1). A guided dissection of a large, simple flower such as a lily (*Lilium spp.*) would provide an opportunity to cover fundamental floral terminology. Leaf, stem, and root terms are also used throughout the text and could be modeled using a "pothos" plant (*Epipremnum spp.*) or other live specimens as available. Students could attempt to assign correct terms to the material first based on prior knowledge and then revise their understanding as they review definitions for each term.

An alternative approach is to have students work together in small teams to discuss their prior experiences with the terms. Students should be familiar with many of these words as Tier 1 science vocabulary terms, but the words may be used in a different context in their everyday language. For example, students may have heard someone say, "The root of the problem is . . ." or ". . . is a thorn in my side!" Encourage students to consider how they have heard these words used and how they may relate (or differ) from their botanical definitions. For terms that are still unclear, students could work together and use other discipline-specific readings to refine their understanding.

Applying Terminology

To encourage further engagement with botanical terms, teachers can have students choose one of several species of plants mentioned in *This Poison Heart* and produce a labeled drawing using terms listed in table 4.1. As most of the plants referenced in the text actually occur in nature, students should be able to locate botanical illustrations for them online. Filtering image search results to include only those that are black and white or grayscale will yield floral or vegetative diagrams that the students can download, print, and label. Students could then exchange these with classmates and take turns scoring the number of terms applied correctly/incorrectly. In cases where students disagree on how a term should be applied, the class as a whole can discuss and consult additional textbooks or other reference materials to aid in their decision.

Table 4.1. Botanical Terms Used in *This Poison Heart* (listed in order of first appearance in text)

Term	First Appearance in Text	Definition
bud	p. 3	an immature shoot that normally occurs in the axil of a leaf or at the tip of a stem
flower (blossom)	p. 3	reproductive organ that consists of stamens and carpels and typically surrounded by petals and sepals
leaflet	p. 3	each of the smaller leaf-like structures that together make up a compound leaf
offshoot	p. 3	a sprout or shoot that branches off from the main stem
petal	p. 3	modified leaves that surround the reproductive parts of flowers; often brightly colored to attract pollinators
pistil	p. 3	the ovule-producing (female) part of a flower; consists of a stigma that receives pollen and a style that leads to an ovary at the base
seed	p. 3	an embryonic plant enclosed in a protective outer covering (seed coat) and a food reserve (endosperm)
stalk	p. 3	the main stem of an herbaceous (non-woody) plant
stem	p. 3	the main axis of a plant, where leaves and branches are attached
thorn	p. 3	a modified, sharp-pointed branch, often used by plants as defense against herbivory
branch	p. 4	a structural extension of a plant, connected to the central stem of a plant, or trunk of a tree or shrub
root	p. 4	the portion of a plant used for the uptake of water and nutrients; typically underground
flora	p. 6	the full assemblage of plants that occur in particular region or habitat
vine	p. 6	the slender stem of a trailing or climbing plant
leaf	p. 10	a flattened structure, typically green and blade-like, attached to a plant stem; primary function is photosynthesis
tendril	p. 10	a slender threadlike appendage of a climbing plant that twines around other plants or structures to give the plant support
botany	p. 11	the scientific study of plants, including their physiology, structure, genetics, ecology, distribution, classification, and economic importance
fruit	p. 13	a structure that encloses and protects the seeds of flowering plants; develops from the ovary at the base of the pistil

(continued)

Table 4.1. (Continued)

Term	First Appearance in Text	Definition
poisonous plant	p. 13	a plant that when touched or ingested in sufficient quantity can be harmful or fatal to animals
leaf teeth	p. 20	tooth-like projections on the margins of many leaves
vein	p. 20	a conductive structure that consists of xylem and phloem; used for transport of water and sugar and for support
pollen	p. 34	a powdery substance present in plants in seed-bearing plant; responsible for production and delivery of male gametes
bark	p. 59	the outermost layers of the stems and roots of woody plants
medicinal plant	p. 90	plants that possess therapeutic properties or exert beneficial pharmacological effects on animals
node	p. 90	the location on a stem where leaves are attached and axillary buds are located
trichome	p. 90	a small hair on the epidermis of a plant, typically unicellular and glandular
foliage	p. 87	the collective term for the leaves on a plant

Plant Awareness Disparity

In recent years, the botanical community has identified a concept called Plant Awareness Disparity (PAD). Originally described as "Plant Blindness" by Wandersee and Shussler (1999) and renamed by Parsley (2020), PAD occurs when people fail to appreciate the abundance and importance of plants in their daily lives. Hershey (1996) proposed that this lack of familiarity with plants has been reinforced during K-12 education because science textbooks systematically favor animal examples over plants. Brownlee et al. (2021) noted that until textbook developers address the problem of PAD, it will be necessary for educators to include additional plant examples to balance the representation of plants and animals in their own classrooms.

The following activity uses student memory probes to help address PAD in students. Drawing on themes students will encounter when reading *The Poison Heart*, students are asked to think back to childhood memories that involve poisonous, painful, mysterious, or magical plants. To begin, the teacher prepares cards with prompts and places them face down on a table. Some suggested prompts include:

- Describe a plant that your family warned you to stay away from as a child. What was dangerous about it?

- Was there a plant that your family used to cure ailments or injuries? What parts were used, and how was it prepared?
- Describe a plant you thought was mysterious or magical as a child. What about the plant made you think it had powers?
- Do you remember any plants that made sounds? Describe the sounds and what you think caused the sounds.
- Did you or anyone you knew ever get injured as a result of a plant? What happened?
- Growing up, do you recall any plants you thought looked like, or acted like, animals?

Students draw cards randomly, read the question aloud, and then share their answers with the class. After the student responds, classmates can comment or add their own memory to enhance the discussion of each prompt. Students may then investigate the poisonous or medicinal properties of the plant they described and explain how and where these chemicals are located in the plant using appropriate botanical terms.

What's in a Name?

While most of Bayron's in-text examples refer to poisonous plants used for malevolent purposes, medieval physicians often utilized these same plants as medicine (Bevan-Jones, 2009). The *Doctrine of Signatures*, an herbal folklore dating back to the fifteenth century, asserts characteristics of plants such as shape or texture indicate the medicinal uses of a plant or their ability to ward off negative spirits. The names of ten species referenced throughout the text by either their Latin or common name are listed in table 4.2.

For this activity, the teacher should provide an overview or review of Linnaeus' binomial nomenclature as it applies to naming species and contrast

Table 4.2. Examples of Plants Referenced in *This Poison Heart*

Latin Name	Common Name	First Appearance in Text
Calendula officinalis	pot marigold	Ch. 1, p. 10
Cicuta douglasii	water hemlock	Ch. 2, p. 18
Abrus precatorius	rosary pea	Ch. 5, p. 60
Aconitum spp.	wolfsbane	Ch. 5, p. 60
Ageratina altissima	snakeroot	Ch. 5, p. 60
Nerium oleander	oleander	Ch. 5, p. 60
Ricinus communis	castor bean	Ch. 5, p. 60
Tacca chantrieri	bat flower	Ch. 8, p. 98
Atropa belladonna	deadly nightshade	Ch. 10, p. 123
Symphytum officinale	comfrey	Ch. 11, p. 140

this with the use of common names. Students may then be assigned one of the species commonly referenced in the text and use a graphic organizer to help guide them through the process (see figure 4.1, for example).

The etymology of common names is especially interesting from a historical perspective; for example, "wort" is part of many common names (e.g., lungwort) and derives from the Old English *"wert,"* meaning plant. Likewise, "bane" (e.g., wolfsbane) comes from *"bana,"* meaning "causing death, or poison." Internet resources that will be useful include Wikipedia (the most user-friendly option; search by scientific or common name and explore the resulting information), Botanary (hosted by Dave's Garden, search by genus or specific epithet to see the meaning of word roots, Latin or Greek derivation, and pronunciation), and World Flora Online (search by scientific name to see images, distribution maps, common names, and more). Stearn's *Botanical Latin* and Mabberly's *The Plant Book* are excellent print references and will likely provide the most information on names.

Once students complete their investigations and graphic organizer, they should create a short oral presentation that includes the following elements: (1) illustration or photo of the plant; (2) translation of the Latin name; (3)

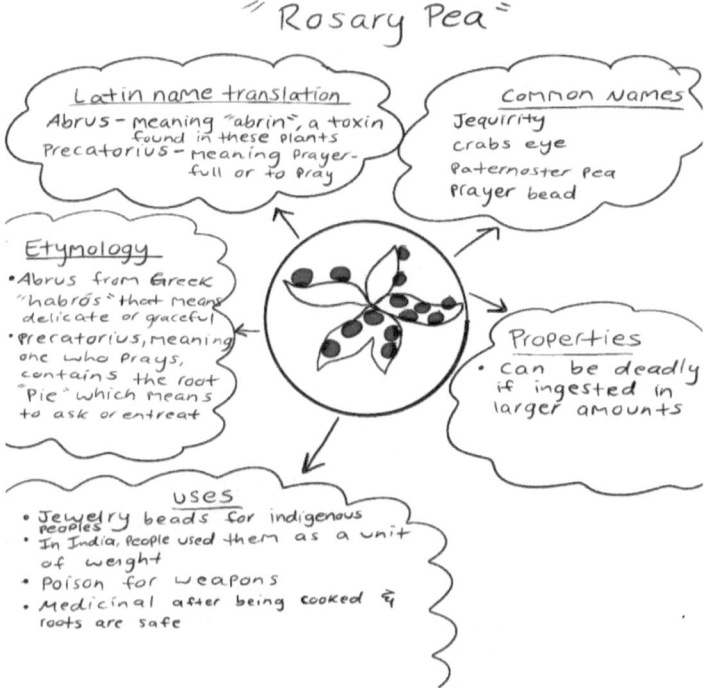

Figure 4.1. Sample Graphic Organizer for Investigation of Plant Etymology (created by authors).

list of common names applied; (4) the etymology of scientific and common names; (5) useful or dangerous properties; and (6) historical and/or modern uses. Bayron maintains a high degree of accuracy with respect to the botanical details of the non-fictional species that appear in the text. As students encounter these plants in the book, this background research will allow them to better understand and connect each plant's significance and relevance to Bri's story.

WHILE READING *THIS POISON HEART*

A central idea in *This Poison Heart* deals with classification of the natural world, both literally and metaphorically. Bri is acutely aware of the plants around her and knows them by name, their curative or poisonous properties, and if they are to be used—or feared. This concept of plant classification is readily connected from the text to the science classroom. For example, the very first sentence in chapter 1 begins with classification: "White roses. Genus *Rosa*. Family Rosaceae. Common name 'Evening Star'" (p. 1). This ties well with the investigation into scientific names described in the "before reading" section but also provides an opportunity to discuss the Linnaean hierarchy (e.g., family names end in -aceae; family is more inclusive than genus). Ethnobotany, or human uses of plants, is another type of classification that occurs repeatedly throughout the text. Bri's home contains an apothecary (a place to obtain natural medicines and sometimes magical potions), with jars of plants lined on shelves from floor to ceiling.

Investigation into the *Absyrtus Heart*

On pages 160–162, Bri discusses some of the jar's contents with another character, Marie. Bri describes *Symphytum officinale* as common comfrey but suggests that *S. uplandicum* would be better for healing ulcers due to its higher alkaloid content (p. 161). Throughout the book, plants are described as "deadly," "poisonous," "medicinal," or as "cures," each with specific requirements to ensure the desired effect. Additionally, the theme of plant vs. animal (human) occurs prominently with the introduction of the *Absyrtus Heart* in chapter 7 (p. 91). Bri discovers a book entitled "*Venenum Hortus*," where each page describes a different species of a poisonous plant. On page 91, a written description of this strange plant is revealed: "rope-like stalks and tufts of black leaves" and "veins running across its fleshy pink surface." The readers will later discover that the *Absyrtus Heart* is a plant that rose up from the human remains of one of Bri's ancestors after burial (p. 301).

As students read *This Poison Heart*, it is important for them to pay close attention to this unique plant. The information they gather will be used to evaluate the potential for a hybrid plant-animal organism to occur in the natural (non-fictional) world. Students may compare and contrast plant and animal cells, investigate cell theory, and research examples of organisms with characteristics that do not fit easily into clear taxonomic categories (e.g., viruses or protozoans). Depending on the information they find, students will make conclusions about the plausibility of the existence of the *Absyrtus Heart*. A four-part activity described below will guide students through the fundamental science content needed to evaluate if this plant could actually occur in nature. One or more of these activities could be repeated periodically as students read the remainder of the text and gather new information or evidence to support their learning.

Part 1. Card-Sort

To begin, students complete a card-sort activity with descriptors of plant and animal cells. These terms include organelles and structures that are present in plant cells, animal cells, or both (see table 4.3). Students should create a card for each term, then sort the terms into one of the three categories: Animal cells, Both, Plant cells.

Part 2. Initial Class Discussion

After the card-sort activity, have students engage in a discussion, either in small teams or the whole class. Refer to page 91 of the text for the descriptions of the *Absyrtus Heart*. Based on this account, draw and label this plant. Then answer the following questions: *If you looked at its cells under the microscope, what do you think you would see? Why?* Draw and label what you would see under the microscope.

Optional: Teachers can provide students access to a microscope and a set of prepared slides including plant and animal specimens. Students can view

Table 4.3. Possible Cellular Terms to Include in Card-Sort

Animal Cells	Both	Plant Cells
Centrioles	Nucleus	Cell wall
	Mitochondria	Chloroplast
	Smooth and rough endoplasmic reticulum	Plasmodesmata
	Ribosomes	
	Vacuole	
	Golgi apparatus	
	Lysosomes	

these slides to refine their idea of how tissue(s) from the *Absyrtus Heart* might appear under the microscope.

Part 3. Collecting Information

Next, students will research cell theory and cell/tissue types through the lens of the *Absyrtus Heart*. At the end of chapter 7, students will be ready for the activity on cell theory since they have been introduced to a detailed description of the *Heart* in the text. Students will be challenged to determine if this description is consistent with scientific principles. The three basic tenets of cell theory are: (1) all living things are composed of one or more cells; (2) the cell is the basic unit of structure and organization in organisms; and (3) cells arise from pre-existing cells. Modern cell theory adds more specific interpretations, including the following: (4) cells contain genetic information in the form of DNA that determines the characteristics and functions of the organism; and (5) in organisms of similar species, all cells are fundamentally the same. Students are typically familiar with plant and animal cells, and they may be familiar with the types of cells observed in various mammal tissues. Histology slides and mitosis/cell type slides such as onion root tips and whitefish blastula could be used to review or instruct on these differences. Students should investigate organisms less common in the typical science curriculum to become familiar with different cells, such as bacteria, protozoa, algae, fungi, or slime molds. This activity will help students build a foundation for critically evaluating the *Absyrtus Heart* when they get to chapter 24, and its origin story is revealed.

Optional: A classroom discussion about the difference between scientific theory and scientific law could be a timely addition to this activity. Because cell theory is a "theory" (a term that sometimes holds a different meaning when used in everyday language), some students may have incorrect preconceptions that it is not held in the same level of factual regard as a scientific law.

Part 4. Class Discussion after Information Gathering

After completing microscope investigations, students can engage in further discussion using the following questions/prompts as a guide:

- In a single living organism, are all cells exactly the same? Give some examples.
- In a complex multicellular organism, describe the commonalities you can find across different types of cells.
- Discuss how an organism knows how to make different types of cells.
- In what ways is it possible for an organism to make cells that are vastly different from the parent cells? What would constitute "vastly" different, in your opinion?

AFTER READING *THIS POISON HEART*

Origin of the *Absyrtus Heart*: Making an Argument

The origin of the *Absyrtus Heart* is described in chapter 24,

> Jason made use of the fleece and took the princess Glauce as his wife, and so became King. Medea and her children retreated to the island of Aeaea, home of the sorceress Circe. She spent her days wandering her Poison Garden, where she buried the six pieces of Absyrtus's body. In the spot where the earth covered each piece of his remains, peculiar plants grew, plants only Absyrtus's beloved family could tend. Medea nurtured them with drops of her own blood and slivers of moonlight. (p. 301)

As the culminating task, ask small teams of students to evaluate the origin of the heart to respond to the question: *Could a plant arise from animal tissue?* Students may use the CER (Claim-Evidence-Reasoning) framework to support their perspective, using evidence derived from their investigations of different cell types, cell theory, and the differences between plant and animal cells. First, students should write a claim statement that answers the question (approximately one sentence in length). Next, students should add two to three examples of evidence from the text that directly support their claim. Finally, students will connect their evidence to reasoning, using background information and science concepts they have learned during their reading. Student teams should present their arguments to the whole class.

Examining Diversity, Equity, and Inclusion in *This Poison Heart*

Discussions of diversity, equity, and inclusion (DEI) may easily be built around concepts found throughout *This Poison Heart*. First, not only is Bri a Black female with a strong interest in biology, but also she has a unique ability to engage with plants. Throughout the text, she frequently poses and tests theories to explain her peculiar interactions with the plant world. For example, Bri describes her interaction with eating yew berries and her friend having to go to the ER, while Bri was unphased by the berries. Bri recalls, "I'd eaten exactly one less than her. But I should have had the same symptoms. I should have felt something" (pp. 13–14). And on page 96, Bri questions, "Contact with the skin caused rashes and pain that could last for days. I didn't want to have to douse myself in calamine lotion, but after my encounter with the poison ivy and hemlock, I wondered if I even needed to worry about it." She recognizes she is different.

Second, her family unit would be regarded as "non-traditional" under historical societal norms, with two moms in a committed lesbian relationship. In addition, Bri's own romantic feelings for Marie give readers an opportunity to find a representation of characters that may be similar to themselves. These examples of human diversity are important to help students develop skills of empathy, inclusion, and acceptance.

The following activity has four parts and addresses DEI both in the text and in students' lives. As students read the book, group together six segments of text (chapters 1–5, 6–10, 11–15, 16–20, 21–25, 25–31). First, for each segment of text, ask students to keep a brief journal of how Bri feels throughout that particular part of the story. Record her emotions, words, actions, reactions, inclinations, and responses to how she's treated and how she feels about what's happening to her family. After each section of text, arrange students in teams of three to four and ask them to share their journal entries with their team. Once shared, have each team discuss the similarities and differences presented and then as a whole team, respond to and discuss the following prompts:

- Describe a few examples in this section of text where Bri felt excluded, judged, and/or isolated because of her race or gender. If you feel comfortable, share with your team when and how a similar experience you may have had made you feel.
- Describe Bri's experiences, both positive and negative, with her STEM identity in this section of text. Include textual evidence to support your description(s).

Fourth and lastly, bring students back together to share team discoveries and takeaways from the DEI discussion.

EXTENSION ACTIVITIES

Nature vs. Nurture

The ongoing debate between nature and nurture is commonplace in the fields of psychology and human development. In this text, nature and nurture is of importance because Bri inherits unique magical abilities from her biological lineage. Her botanical abilities are described in detail in chapter 1 and include reviving dead plants, making plants double in size almost instantly, and being able to handle extremely poisonous plants without injury. Although Bri was raised by adoptive mothers, she retains her ancestral abilities. Within the realm of science, the idea of nature vs. nurture is important because traits are passed to offspring through genetic means, but offspring also learn behaviors from parents. Using Bri's unique ability as a starting point, what are some other traits that can be classified as behavioral vs. inherited in humans or other animals? What other abilities do humans have that come from their nature, nurture, or a combination of both? Students could choose or be assigned an ability and explain how both nature and nurture contribute.

Plant Chemistry

After finishing *This Poison Heart*, students are likely to find plant poisons an intriguing topic for further investigation. What compounds make poisonous plants so toxic? Students may explore the chemistry and ecology of defensive plant secondary metabolites (e.g., phenolics, glycosides, alkaloids, and terpenes). Students could conduct independent experiments using monarch butterfly caterpillars and their host plants in the milkweed genus (*Asclepias spp.*), which are filled with cardiac glycosides. What feeding behaviors do the caterpillars use to avoid poisoning from milkweeds? Are there certain species of milkweed that caterpillars prefer based on their glycoside content?

Alternatively, teachers can have students ask these questions using data collected as part of large-scale citizen science projects. For example, CitizenScience.gov hosts a monarch larva monitoring project where participants learn how to identify milkweed plants, monarch butterfly eggs, caterpillars, and pupae. Subsequently, they record the number of eggs and caterpillars present on the plants they are monitoring over several weeks. Students can examine volunteer observations to see if the numbers of larvae increase or decrease in relation to their host plants. All data collected as part of the monitoring project are publicly available and can be downloaded for students to work with in the classroom.

In Their Own Words

Students who have not experienced the types of challenges faced by underrepresented or marginalized groups of people may be unable to relate to others who have. One way to help students become more empathetic is to have them learn about other people's experiences in a meaningful way. Additionally, diverse students need role models and mentors that look like them so they can visualize themselves in the same role. For this activity, students should select a biologist, living or historical, and tell the story of the biologist's life and career in the first person. Students should conduct research and/or interviews to gain a better understanding of the challenges that individual has faced. Students should consider scientists who are members of any of the following communities: traditionally underrepresented cultural or ethnic groups (e.g., Black, Latinx, and Indigenous peoples), people with disabilities, or LGBTQIA+. Fortunately, there are now many resources available online that highlight the stories of scientists in these categories, including the National Institutes of Health, the National Science Foundation, Science Advances, Black in Natural History Museums, and Pride in STEM. Students should choose someone with whom they are not already familiar. The main goal is for students to gain meaningful insights into the challenges and victories the individuals they choose have experienced on their path to becoming scientists. Sample interview questions listed below are designed to help students better understand the life of a scientist and develop empathy.

- Please describe your general research focus and any specific interests you have in your field.
- Describe a typical day at work. What types of activities do you engage in?
- Tell me about educational preparation. What was your major? Where did you go to school? How did your education prepare you for your career?
- Describe several of your greatest accomplishments.
- Who, or what, had the most significant influence on your career, and why?
- Have you ever felt excluded or alone as a biologist from a diverse background? Why or why not?
- What differences do you see in the challenges students face today compared to when you were a student?
- Do you have any advice for students like me? How can I learn from your experience?

Students could prepare a graphic organizer (see figure 4.2) that highlights the career path and the ups and downs experienced by the scientist they interviewed. Each student can deliver a presentation to the class. Common themes will likely arise in these presentations, and the activity can end with students writing a self-reflection on what they have learned about themselves and the scientists in the process.

Botanical History

Throughout *This Poison Heart*, the reader learns that Bri's family has a magical and dangerous botanical history. Some students may become interested in their own family's relationship to plants after finishing the text. Going back generations, what is their unique botanical "family tree?" Students could interview family members to find out which plants are meaningful to them and/or were important to their ancestors. Examples of significant plants could include those that have agricultural or food uses, relate to holidays or

Figure 4.2. Sample Graphic Organizer for in Their Own Words (content from Kimmerer, 2020).

religious ceremonies, hold sentimental meaning, provide economic benefits, or contribute to their livelihood in a different way.

One way to document botanical history is through herbaria, which are museum collections of dried, pressed plants collected, preserved, and studied by scientists. Students could create their own herbarium specimen and explore the digitized specimen information available from herbaria around the world. One source of materials for engaging in classroom learning about botanical history and herbaria is *Rooting Your Students in Their Botanical History*, searchable online, which provides instructional modules for teachers.

CONCLUSION

Through the story of Briseis, students are encouraged to be more conscientious of the huge variety of plants that surround them in the natural world. Readers have multiple opportunities to make connections to the science content of botany through the reading of *This Poison Heart*, from introductory biological concepts like binomial nomenclature and cell structure to more advanced topics such as plant chemistry and DEI. Based on the classroom style, students may engage and demonstrate understanding via whole-class instruction or individual projects through the variety of activities presented in this chapter including graphic organizers, card-sorts, class discussions, and interviews. Most importantly, these activities highlight plants as an accessible, yet frequently overlooked opportunity for scientific learning and investigation.

This Poison Heart offers the reader a suspenseful story and a diverse suite of characters with unique backgrounds, relationships, interests, and abilities. The science content related to botany—including plant classification, plant chemistry, cell theory, ethnobotany, and STEM identity—becomes engaging for students who may not have previously realized an interest in the subject. The focus on diverse populations reminds us that everyone has the potential to make important contributions to the field of science, and a diversity of perspectives is beneficial when engaging with natural phenomena through the process of sense-making.

REFERENCES

Bar-On, Y. M., Phillips, R., & Milo, R. (2018). The biomass distribution on Earth. *Proceedings of the National Academy of Sciences, 115*(25), 6506–6511.

Bayron, K. (2021). *This poison heart*. Bloomsbury.

Bevan-Jones, R. (2009). *Poisonous plants: A cultural and social history*. Windgather Press.

Boswell, H. C., & Seegmiller, T. (2016). Reading fiction in biology class to enhance scientific literacy. *The American Biology Teacher, 78*(8), 644–650.

Brownlee, K., Parsley, K., & Sabel, S. (2021). An analysis of plant awareness disparity within introductory biology textbook images. *Journal of Biological Education*. Advance online publication. https://doi.org/10.1080/00219266.2021.1920301.

Hershey, D. R. (1996). A historical perspective on problems in botany teaching. *The American Biology Teacher, 58*(6), 340–347.

Kimmerer, R. W. (2020). *About*. https://www.robinwallkimmerer.com/about.

Knapp, S. (2019). Are humans really blind to plants? *Plants, People, Planet, 1*(3), 164–168.

National Research Council. (2012). *A framework for K-12 science education: Practices, crosscutting concepts, and core ideas*. National Academies Press.

Parsley, K. M. (2020). Plant awareness disparity: A case for renaming plant blindness. *Plants, People, Planet, 2*(6), 598–601.

Wandersee, J. H., & Schussler, E. E. (1999). Preventing plant blindness. *The American Biology Teacher, 61*(2), 82–86.

Chapter 5

Exploring Nature and the Nature of Scientific Inquiry

Reading The Evolution of Calpurnia Tate

Amy Palmeri, Emily Pendergrass, and Heather Johnson

Despite pervasive socio-ecological challenges in the world, children today do most of their learning about science in school, disconnected from the world they live in. They are more likely to be able to name more Pokémon characters than be able to classify real organisms in the natural world. This disconnection from nature has a profound impact on children—including a rise in obesity, attention disorders, and depression—and extends beyond childhood shaping many of our behaviors into adulthood. For example, adults who connect to nature are more likely to use fewer natural resources—like electricity or taking fewer and shorter showers and increase our willingness to engage in pro-environmental behaviors.

The good news is that connecting students to nature does not require sending students out into the wilderness. No matter the environment, nature is all around. Teachers and students can explore the schoolyard, their neighborhoods, local parks, and other public spaces that they spend time in daily. Even in urban spaces, there are many things we might notice if we set the intention to notice them: trees, birds, squirrels, small weeds growing in the sidewalk cracks, and flowers planted in front yards or in window boxes. Even though this form of nature is highly cultivated and maintained by humans, it is nature nonetheless. Inquiry into the novel *The Evolution of Calpurnia Tate* (Kelly, 2011) is aimed toward engaging students with nature in positive ways—both so they learn more about interactions within ecosystems and also that they are inspired to be good stewards of the environment around them.

THE EVOLUTION OF CALPURNIA TATE BY JACQUELINE KELLY

The Evolution of Calpurnia Tate is a young adult historical fiction novel that tells the story of eleven-year-old Calpurnia (Callie) and her desire to become a naturalist. Set in the small farming community of Fentress, Texas at the turn of the twentieth century, Callie confronts the realities of the societal forces and familial expectations that shape her future. As the middle child and the only daughter in a family of seven children, Callie much prefers spending time exploring and questioning the natural world around her than she does engaging in household pursuits more typical of girls during this time period. Thus, the concept of evolution in the text is as much about Calpurnia's personal growth and evolution as it is about what she comes to learn about the workings of the natural world. The text provides rich descriptions of the interdependence of organisms within an ecosystem and draws on the explanatory power of the relatively new and still controversial Darwin's Theory of Evolution (the historical fiction is set in 1899 and *On the Origin of Species* was published in 1859).

Note to Readers

The science content foregrounded in this chapter is the nature of scientific discovery and the related science practices that characterize how new scientific knowledge is constructed, communicated, and critiqued. The before-, during-, and after-reading strategies suggested throughout will help students engage with the text as they develop, apply, and refine the ability to employ the eight Next Generation Science Standards (table 5.1) and engineering practices.

Place-Based Education Tasks and Connection

Many of the tasks we've designed to use with this novel center on the overarching idea that there is a power that comes with focusing on local places. Pragmatically, they are easy to access, but place-based educators argue that local places allow teachers to turn communities into classrooms (Place-Based Education Evaluation Collaborative, 2010). By fostering students' connection to local places, they come to understand the intricacies of the world around them. These local places serve as a microcosm of the broader world and create rich opportunities for teachers to situate the teaching of science concepts. A focus on the familiar increases the relevance of the science concepts being learned and thus has the potential to generate, in students, both a desire and sense of agency to care for the world around them. The Place-Based Education

Table 5.1. Science and Engineering Practices

Scientific and Engineering Practices	Explanation
Asking questions (for science) and defining problems (for engineering)	To determine what is known, what has yet to be satisfactorily explained, and what problems need to be solved.
Planning and carrying out controlled investigations	To determine what data to collect that can be used to test existing theories and explanations, revise and develop new theories and explanations, or assess the effectiveness, efficiency, and durability of designs under various conditions.
Analyzing and interpreting data	With appropriate data presentation (graph, table, statistics, etc.), identifying sources of error and the degree of certainty. Data analysis is used to derive meaning or evaluate solutions.
Using mathematics and computational thinking	As tools to represent variables and their relationships in models, simulations, and data analysis in order to make and test predictions.
Constructing explanations and designing solutions	To explain phenomena or solve problems.
Engaging in argument from evidence	To identify strengths and weaknesses in a line of reasoning, to identify best explanations, to resolve problems, and to identify best solutions.
Obtaining, evaluating, and communicating information	To share ideas and findings with others—orally, in writing, through representations, or in discussions with others. This also includes being able to interpret scientific texts in order to derive meaning, evaluate validity, and integrate information into scientific explanations.

Source: National Research Council, 2012.

Evaluation Collaborative (2010) found that such an approach supports the following positive outcomes: (1) improves student learning; (2) invites students to become active citizens; (3) energizes teachers; (4) connects schools and communities; and (5) encourages students to become environmental stewards. In short, through building an awareness of the local environment students can also engage in thinking at the global level. While taking global action feels daunting, local action, by nature of being "close to home" is highly relevant and personal. This, in turn, nurtures a sense of agency in students as they are more likely to see evidence of the positive impact of their actions which are informed by their deep understanding of related scientific concepts. The following tasks have been designed to support growing understanding of science concepts and practices in the given place where the students are reading. We will share a sampling of ideas to use before, during, and after reading the novel, as well as ideas to extend learning beyond the text.

PREPARING STUDENTS TO READ *THE EVOLUTION OF CALPURNIA TATE*

Scientific Observation

Before introducing *The Evolution of Calpurnia Tate*, teachers may want to engage students with an intentionally designed observation task requiring them to grapple with what constitutes scientific observation (in contrast with everyday noticing or inferences). Systematic, scientific noticing and observations are perhaps the most fundamental of all science practices. We tend to wonder about things we notice that are unexpected or when a unique detail has caught our eye. Such wonderment often compels us to ask questions (an example of questions and wonderment from Calpurnia can be found on p. 117). It is the pursuit of answers to such questions that leads to the authentic application and refinement of the science practices, such as *asking questions* about phenomenon or *planning investigations* to *construct explanations* about causal mechanisms behind observed phenomenon.

For this task, we recommend that the teacher collects a set of objects that share features but that also have variety within a type. We have found that a collection of lemons, limes, and clementines work well as these fruits are similar in size and texture, but each individual lemon, for example, will have variations that make it distinct from the other lemons included in the set. When launching this task, the teacher should give each student one item—in this case one lemon, lime, or clementine—and a blank sheet of paper. The teacher then simply asks students to observe their piece of fruit and record their observations on the paper. Teachers need to intentionally keep the directions vague as we want students to write their initial observations, which will likely not be specific. Students could record anything they notice, such as size, color, and bumps on the fruit. After students record their observations, collect the fruit and place them in an opaque container (this serves to mix the fruits up a bit and remove them from sight). Then place the collection of fruit in front of the students. Challenge each student to find their original piece of fruit.

This before-reading activity is most successful when the ultimate purpose of the observation is hidden from the students. As the teacher you are hoping students' initial observations are superficial, thus making it difficult for them to identify their original piece of fruit. It is this challenge, more than any other, which allows students to experience how everyday observations differ significantly from the more systematic scientific observations—both in terms of what we notice and how we might record details of our noticing. Scientific observation specifically relates to observations using the senses (i.e., touch, sight, and hearing) or measurement to gather information.

After students realize that they are unable to find their original piece of fruit, they can engage in a second round of observations. To support more systematic observations, the teacher could introduce tools (tape measures, scales, hand lenses, etc.) that would facilitate more precise observations and thus more precise recorded notes. Additionally, to distinguish an individual piece of fruit, students can answer questions such as: is there a unique mark on your piece of fruit? If so, what is the shape, color, and size of the mark? Given the nature of this activity, a teacher may want to use this at the beginning of the school year and leverage it anytime careful scientific observations are critical. There are many connections that a teacher could make to this before-reading strategy during the actual reading of the text. For example, early in the text Calpurnia is proud of the observations she has recorded in her "scientific notebook" but then has her feelings hurt when her Granddaddy tells her that her recorded observations are only a "fair start" (pp. 24–25). As students continue reading the text they might be asked to consider how Calpurnia's scientific noticing and recordings in her "scientific notebook" become more sophisticated throughout the text.

In conjunction with and connecting to students learning about scientific observation, students can record their observations in a personalized "Evolution of *student's name*" notebook (strategy elaborated on below). Teachers can choose excerpts from *Citizen Scientists: Be a Part of Scientific Discovery from Your Own Backyard* (Burns & Harasimowicz, 2013) to reinforce the notion that there is much for students to explore and discover about nature right in their own backyards. Additionally, we recommend using videos from John Muir Laws about Nature Observations. For example, Episode 8 on Asking Questions (Laws, 2017) is particularly relevant as it asks students to inquire across different variables of the same species (i.e., observation of a bird: flight pattern, call, color, and diet). After watching all or selected clips of this video, students can then go outside, spread out across a natural area, and observe and practice.

First Lines

Typically, the First Lines before-reading strategy requires students to read the first few sentences of a text they will be working with (AdLit.org, n.d.). Students consider these first lines as they are asked to make predictions about what the text will be about. This strategy is particularly useful as a means of activating students' prior knowledge and as a way to scaffold students' comprehension of the text.

When implementing the First Lines strategy with *The Evolution of Calpurnia Tate*, we recommend that rather than choose the first lines of the book itself, the teacher choose several of the quotes, taken directly from Darwin's *The Origin of Species*, that Jacqueline Kelly includes at the beginning of each chapter. Each quote is rich with the potential to activate students' prior knowledge and introduce them to domain-specific vocabulary or scientific processes. However, because students will be working with direct quotes from a primary-source text published in 1859, teachers may want to have students work collaboratively in small teams rather than individually. When selecting quotes to use with this strategy, teachers should consider the specific reasons they have chosen to use this book with their student scientists. To facilitate this process, we have selected a few quotes from *Calpurnia Tate* that could be used productively with the First Lines strategy (table 5.2). We justify the richness of each quote by identifying relevant science practices, disciplinary core ideas, and cross-cutting concepts from the *Next Generation Science Standards* (NGSS Lead States, 2013) that a teacher could leverage when using this text in the secondary science classroom.

The teacher can begin by writing the chosen quotes on an index card—choosing one quote for each small team of students. To provide context for these quotes and the task, a teacher might say: *This is a story about a young girl named Calpurnia, who wants to become a naturalist. You will be given some quotes from the book and your job will be to make sense of the quote as you consider what the quote might be telling you about what a naturalist does.* With this context established, the teacher previews the First Lines strategy graphic organizer (see figure 5.1). Our graphic organizer and directions for the activity are a variation of the typical First Lines strategy and better reflect our goal that students identify and engage with domain-specific vocabulary and/or science practices mentioned or implied in the quote. In teams, students should: (1) begin by recording their quote on their graphic organizer; (2) list any science process skills named or inferred in the quote; (3) list any domain-specific words or phrases included in the quote; discuss and record tentative ideas regarding the meaning; and (4) discuss what they think the quote means and make a prediction about how the text might provide insight regarding the question: *What does a naturalist do?* Following this small-team discussion, the teacher can engage students in a brief discussion of the domain-specific vocabulary and the students' predictions about what a naturalist is and what a naturalist does. Later in the unit students should return to the predictions they made while studying the first lines to make modifications to their understanding.

While Reading *The Evolution of Calpurnia Tate*

Once students begin reading the text, it is essential that the teacher also develop and use during-reading strategies. Such instructional activities

Table 5.2. Quotes for First Lines Strategy

First Lines from The Evolution of Calpurnia Tate	Chapter and Page	Relevance to …
"When a young naturalist commences the study of a group of organisms quite unknown to him, he is at first much perplexed to determine what differences to consider … for he knows nothing of the amount and kind of variation to which the group is subject … "	Ch. 1 p. 1	**Science Practice(s):** Asking questions; Developing and using models **Disciplinary Core Idea(s):** Inheritance of traits; Variation of traits **Cross-Cutting Concept(s):** Cause and effect; Structure and function
"We have seen that man by selection can certainly produce great results, and can adapt organic beings to his own uses … But Natural Selection … is a power incessantly ready for action, and is as immeasurable superior to man's feeble efforts, as the works of Nature are to those of Art."	Ch. 5 p. 54	**Science Practice(s):** Analyzing and interpreting data; Using mathematics and computational thinking; Constructing explanations **Disciplinary Core Idea(s):** Natural Selection; Adaptation **Cross-Cutting Concept(s):** Patterns; Cause and effect
"The crust of the earth is a vast museum … "	Ch. 8 p. 94	**Science Practice(s):** Developing and using models; Analyzing and interpreting data **Disciplinary Core Idea(s):** The history of planet Earth; ESS2.B Plate tectonics and large-scale system interactions **Cross-Cutting Concept(s):** Patterns; Scale, proportion, and quantity; Systems and system models
"The action of climate seems at first sight to be quite independent of the struggle for existence; but in so far as climate chiefly acts in reducing food, it brings on the most severe struggle between the individuals."	Ch. 28 p. 334	**Science Practice(s):** Developing and using models; Analyzing and interpreting data; Constructing explanations; Engaging in argument from evidence **Disciplinary Core Idea(s):** Interdependent relationships in ecosystems; Ecosystem dynamics, functioning, and resilience **Cross-Cutting Concept(s):** Patterns; Cause and effect; Energy and matter; Stability and change

The Evolution of Calpurnia Tate
First Lines Graphic Organizer

Team Members: Amy, Emily, Heather

Quote:

"When a young naturalist commences the study of a group of organisms quite unknown to him, he is at first much perplexed to determine what differences to consider . . . for he knows nothing of the amount and kind of variation to which the group is subject" (p. 1).

Important Words and Phrases (with potential meaning)

- organisms: something that is alive
- perplex: puzzling, confusing
- variation: difference

Prediction: What might this tell us about what a naturalist is and what a naturalist does?

A naturalist is a person that finds living things confusing and studies the differences.

_____stop here_____.

We will return to this last section later in the unit.

Checking my Prediction: What is a naturalist? What do they do? How do you know?

A naturalist is a specific type of scientist that studies things in nature through observation and processes. Like a field biologist or ecologist. Callie's brother gives her a notebook for her scientific observations, (p. 8) and she and her grandfather have a multi-day, lengthy discussion about caterpillars (pp. 108-116). See also Callie's observation of the animal in the jar (p. 98).

Figure 5.1. Example First Lines Graphic Organizer (created by authors).

should support both the reading of the text and be designed to support the overarching subject matter learning goals.

Paired Texts: Layering in Non-Fictional Readings and Other Instructional Materials

As a work of historical fiction, the science content embedded in *The Evolution of Calpurnia Tate* is not the primary focus of the narrative. Further, as historical fiction, set at the turn of the twentieth century, students reading the text may assume that the fundamentals of scientific discovery have changed and that the story has little application to the work of science today. The strategy of pairing nonfiction texts and other media with the reading of *The Evolution of Calpurnia Tate* is an effective strategy for engaging students more explicitly with the science subject matter embedded in this fictional text. Pairing nonfiction texts supports students' literacy and science learning by providing scientific content essential for understanding the key ideas from the main text and encouraging students to look at the main text in more rigorous ways.

Exploring Nature and the Nature of Scientific Inquiry

We provide an example of how nonfiction resources can be paired with the reading of *The Evolution of Calpurnia Tate* to support and enhance student learning. Teachers may utilize examples of current discoveries, such as *Scientists Discover the First True Millipede* by Lambert (2022), to read with their students. This article provides opportunities for students to compare the processes of new species identification in 2022 with the process Callie and her grandfather used during the novel set in 1899 (i.e., from the novel: new species found on p. 109; photographs taken chapter 12, letter mailed to Smithsonian p. 177). Students might also examine the process of scientific naming to determine if this new millipede was named after the individual(s) who discovered it the way the new plant (*Vicia Tateii*) was named after Calpurnia and her grandfather (pp. 320–321). For a sampling of additional ideas and pairings, see table 5.3. We urge teachers to follow the leads of their students' interests to help determine what topics to bring alongside reading and learning in this novel.

Naturalist Notebooks and *I See, I Think, I Wonder*

Early in the text, the reader is introduced to Callie's pocket-sized red leather notebook given to her by her eldest brother whom she idolizes. Harry tells Callie, "You can use it to write down your scientific observations. You're a regular

Table 5.3. Nonfiction Text Connections for Knowledge Building

Nonfiction Text	Connection to Novel
Brazil, R. (2020, July 14). Fighting flat-Earth theory. *Physics World*.	On page 13, Calpurnia wonders if she can borrow a book from The Flat Earth Society in San Antonio.
Valverde, J. P., & Schielzeth, H. (2015). What triggers color change? Effects of background color and temperature on the development of an alpine grasshopper. *BMC evolutionary biology, 15*, 168.	Calpurnia contemplates the differences in the colors and sizes of the grasshoppers near her house (pp. 16–17). She questions how the weather and predators play a role in survival.
Bennet, H. (2018, August 13). *From Folklore to Pharmacy*. Chemistry World.	Grandfather and Calpurnia discuss a species they've found. He says, "Here are some nice specimens of sangre de drago, or dragon's blood. The Indians used it to treat gum inflammation" (p.158).
Excerpts from Oberhauser, K. (2015). *Monarchs in a Changing World: Biology and Conservation of an Iconic Butterfly*. Comstock Publishing Associates.	On page 213, Calpurnia captures a spotted fritillary butterfly and preserves it by mounting it.

naturalist in the making" (p. 8). Although Callie wonders what exactly a naturalist is and does, she nonetheless commits herself to spend the rest of her summer being one. She is determined to begin by using what would become her beloved science notebook, to write about the things she sees in the world around her.

For this during-reading strategy, the teacher could give each student their own naturalist notebook, or students can create a digital notebook or even design their own paper notebook. Using a variation of the *I See, I Think, I Wonder* strategy (Ritchhart et al., 2011, pp. 55–63), the teacher can begin much as Callie did in the text, by sending students outdoors either near their homes or in the schoolyard to observe and record. This strategy is particularly relevant to developing science practices because it scaffolds students' ability to use their senses to make careful, specific observations of the things they notice (not just the things they see), to make inferences or conjectures based on what they notice, and stimulates curiosity. It also engages students in thinking about crosscutting concepts as they focus on patterns across observations or consider the relationship between structure and function or cause and effect as they attend to details within their observations. Teachers can also remind students of the pre-reading fruit description task and the need to be specific and detailed in their observations.

When first launching the naturalist notebooks using the *I See, I Think, I Wonder* strategy, students can work through the graphic organizer using one or more of Callie's observations, thoughts, and wonderings as a way of modeling what students will be doing and recording. Following this orientation to the graphic organizer, the teacher can take their class outdoors and ask students to find a place to sit in the schoolyard away from other students. Using a basic *I See, I Think, I Wonder* graphic organizer (table 5.4), students should be given an open-ended prompt to spend five minutes making a list of all the things they can see, hear, smell, or touch from their vantage point. Then the teacher will prompt students to take five minutes to reflect on what they think about what they have noticed. Finally, the teacher can ask students to generate a few *I Wonder* statements.

This strategy is particularly robust as the prompting questions can become more complex or focused as students spend more time observing the same place repeatedly. For example, as students learn more about the living components of the shared place they can be prompted to notice (see) variation within a group of organisms within the environment. Then, they can be prompted to think about how these organisms interact with other organisms in the environment. Finally, students wonderings can become research projects that allow them to more closely examine the conjectures made about the interactions they posit, providing opportunities to develop their scientific practices of asking questions, planning investigations, constructing explanations, and modeling phenomena.

Table 5.4. Example I See, I Think, I Wonder Graphic Organizer

Callie's Noticing	Callie's Thinking	Callie's Wonderings
What does Callie see? Hear? Smell? Touch?	What is she thinking about while observing?	What does she wonder?
See: big, fuzzy caterpillar; 2 inches long; covered in dense fur (pp. 108–109) See: Two black dots buried deep in fur (p. 114)	That's the longest caterpillar I've ever seen (p. 108); Fur looks like a cat's pelt (plush and soft) (p. 109); Don't touch him that hurts (p. 109); Those must be his eyes. He doesn't seem to have eyelids (p. 114).	I wonder what I should feed him? (p. 113); Can I pay my brother to touch him and tell me how he feels? (p. 109) "Why don't caterpillars have eyelids?" . . . "Do caterpillars come as male or female? Or do they turn into male or female while they're asleep in their cocoons?" (p. 114).
See	**Think**	**Wonder**
What do **YOU** see? Hear? Smell? Touch?	What do **YOU** think is going on?	What does it make **YOU** wonder?
See: green bushes low to ground (smells like peppermint gum) hear: hum of insects	I think that maybe the cafeteria workers planted these bushes to use in the cafeteria. I think that it might be crickets or grasshoppers. And maybe bees or something that flaps wings fast.	I wonder if this is what gives peppermint gum flavor? What else is mint used for? Why is it here in the schoolyard? How do scientists tell the difference between insect sounds? Do all bees sting?

Road Map for Reading the Text

Discussion of the text as it is being read, whether in small teams or as a whole class, will support students' comprehension of the text as well as reinforce the connections to scientific practices and relevant science concepts. Thinking across the novel, we've outlined some potential questions for teachers and students to consider as they read (table 5.5).

Create a Graphic Novel of Callie's Journey and Discoveries

A final activity involves teachers pairing the novel with an autobiography of Charles Darwin. We particularly recommend a graphic novel version of Darwin's own autobiography: *What Darwin Saw: The Journey That Changed the World* by Schanzer (2009). Students can learn the terms (i.e., gutter, panel) and tools that graphic novelists use while creating a graphic novel, and then they can work in teams to create graphic novel pages of Callie's journey and discoveries.

Table 5.5. Guiding Questions to Engage Students with Science Concepts Embedded in the Text

Guiding Question	Science Connections
Calpurnia is interested in biology concepts and frequently goes with her grandfather on "nature expeditions." On one such occasion he asks her about the scientific method (p. 28). Discuss why the scientific method is important in science. In what ways can we apply these concepts of experimental design to life's experiences?	**Science Practice(s):** Asking questions; Planning and carrying out investigations
Grandfather is interested in distilling processes and engages in multiple experiments to perfect a recipe (p. 54). Discuss what type of chemical change occurs during distillation. Is this a physical or chemical change and why?	Matter and its Interactions: Analyze and interpret data on the properties of substances before and after the substances interact to determine if a chemical reaction has occurred.
A common treatment for ailments during this time period was baking soda (pp. 79–80). Discuss what baking soda was used to cure in the 1890s–1900s. Is it still used today for medicinal purposes and if so, in what ways?	Matter and its Interactions: Gather and make sense of information to describe that synthetic materials come from natural resources and impact society.
Grandfather said, "Plato said all science begins with astonishment" (p. 105). How does this astonishment connect to Callie's first experience with the microscope?	**Science Practice(s):** Developing and using models; Planning and carrying out investigations.
A dominant theme throughout the novel is the empowerment of women to practice science (see p. 257 for one connection). Describe and discuss some ways women were engaging in scientific pursuits in the novel. What else do we learn about the development of women's rights through reading this text? Women scientists mentioned in the novel (p. 256): • Marie Curie: discovery of radium • Martha Ann Maxwell: naturalist, taxidermist • Mary Anning: fossils • Sofya Kovalevsky: partial differential equations • Lucy Bird: pioneering eco-natural historian	Highlight women associated with different concepts: • Rosalind Franklin: Contributed to the discovery of the molecular structure of DNA • Rachel Carson: concerned about environmental pollution and the natural history of the sea • Maria Merian: accurate illustrator of insects and plants
Callie states, "Creatures sometimes had to die to advance knowledge" (p. 264). Discuss why scientists use animals for research. Propose other alternatives. Discuss how Callie feels about using animals to advance science.	Biological Evolution: Unity and Diversity: Analyze and interpret data for patterns in the fossil record that document the existence, diversity, extinction, and change of life forms throughout the history of life on Earth under the assumption that natural laws operate today as in the past.

(continued)

Table 5.5. (Continued)

Guiding Question	Science Connections
In the final chapter it snows, which is rare in Texas. Callie described the snowy world as quiet and that it lacked living things as "beautiful and menacing" (p. 335). What do you think she means by this? In what ways does snow contribute to the observations she makes about birds, coyotes, and other living things?	Earth's Systems: Develop and use a model to describe how unequal heating and rotation of the Earth cause patterns of atmospheric and oceanic circulation that determine regional climates. Biological Evolution: Unity and Diversity: Use mathematical representations to support explanations of how natural selection may lead to increases and decreases of specific traits in populations over time.

The pages can be combined to make a graphic novel version of the text. These graphic novel pages can be designed using paper and pencil or through design websites like Canva or Mashable. Additionally, these pages can be created at the end of each chapter or discovery to support comprehension of the key ideas.

AFTER READING *THE EVOLUTION OF CALPURNIA TATE*

The after-reading activity proposed in this section leverages the use of the naturalist notebooks created and used by students as they read the novel; however, teachers can opt to engage in this task even if they did not have students create a naturalist journal while reading the text.

Mapping the Schoolyard

Throughout the novel, Calpurnia works to carefully log her adventures with her grandfather (i.e., p. 8) much like the personalized naturalist notebooks mentioned earlier. This practice of carefully noting where and what she finds each day is important especially when they find a new species of a plant. Calpurnia did not map or catalog where she found the plant and had to retrace/recreate her steps to find the exact location (p. 163). Thus, this problem creates the need to map the schoolyard.

Maps are created for many reasons so a first step would be to brainstorm with students why we might want a map of the schoolyard. These might include: *Who is the map for? How will we use the map? What level of detail should we include in the map?* We see this map as a task to create a resource of the schoolyard (field site) for future observations. This natural laboratory will be an investigation into all the biotic (living) and abiotic (nonliving) natural items in the yard.

One way this map project could be done is to create grids for teams to sketch and work within a section of the yard to detail the natural items. A pair of students working in each grid might list and observe all features within their grid. Calpurnia says, "The lack of living things made the landscape both beautiful and menacing" (p. 336), which could open the discussion of abiotic and biotic observations. One student will observe abiotic (nonliving) and the other biotic (living); listing each item, describing it, and observing to see if it is a habitat for a living species (figure 5.2). Then students swap lists and add to the details from their partner. All students are looking out for safety risks such as holes, debris, and so on. Additional tasks that could be completed are: (1) creating symbols that represent the different observed natural items and creating a map legend; (2) measuring boundaries of the yard and distance between items, so that the map can be scaled correctly; (3) using apps and or field guides students can identify the specific species found in the yard.

Directions: You and your partner will be observing biotic (living) and abiotic (non-living) items in your assigned area. Please add any additional notes that would be helpful for yourself and others. Partner 1 completes Abiotic (gray columns) and Partner 2 completes Biotic (white columns). Then swap and add to each other's lists and notes. Be as specific as possible using field guides as necessary.

Grid # 4_____
Partner Names

Abiotic Items	Notes on Abiotic	Biotic Items	Notes on Biotic
rocks		Mushroom	Not sure type (small, white, growing under Oak)
Spider web	From low tree branch to trunk	Oak Tree	
Gum wrapper		butterfly	Orange, blue dots on black wings
Dead leaves		grass	
Rose art crayon		acorn	
feather	robin?	Oak leaves	

Figure 5.2. Example Biotic and Abiotic Note-Taking (created by authors).

Exploring Nature and the Nature of Scientific Inquiry 91

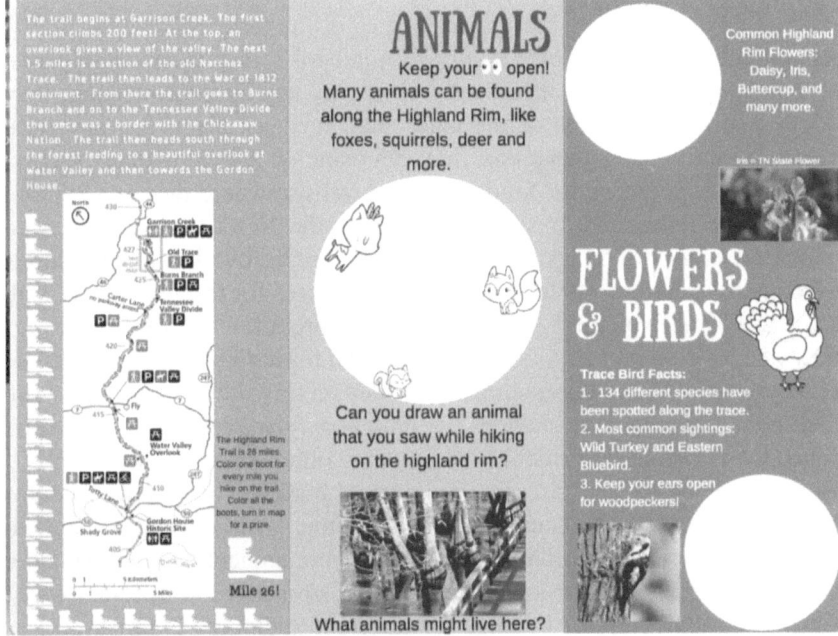

Figure 5.3. Example Eighth-Grade Student-Created Children's Guide to One Section of the Natchez Trace Park in TN. *Source*: Designed using canva.com.

EXTENSION ACTIVITIES

Create a Children's Guide

Students could further use their Mapping the Schoolyard to work with local naturalists to discuss mapping projects at local parks. They can use the new information from the naturalists to create a guide for children to use in learning to observe a specific area (figure 5.3). Students could use apps/websites, field guides, and information identified while mapping to create a guide for students that is specific to the schoolyard and has younger students work with abiotic and biotic items. Perhaps this guide could be multi-paged and each page focuses on a different species found in the yard. Some potential apps/websites that students might use to publish a children's guide are Canva, smore, AdobeSpark, and/or GoogleDocs Templates. National Geographic also has a debris tracker where students can be citizen scientists recording the waste they find in their environments on their website.

"Explainer" Creator

As students create their own naturalist notebooks, map their schoolyards, or simply spend time systematically observing any outdoor location, they might find some element of the place that intrigues them. As they raise questions and seek explanations, they can examine the "Explainer" section of the Scientific News for Students (n.d.) website where they can explore more of the things that intrigue them. Scientific News for Students is an online repository of STEM news and intriguing discoveries; the website is a free multi-genre resource with readability scores for teachers to use with students. The "Explainer" section of this site provides "background pieces for a deeper dive into a broad range of pivotal topics" that students can use as an example. Using the examples provided on this site, students can write their own explainer that includes not only an explanation for the scientific concept but also design sidebars and links to other sources. These student-written explainers can be compiled into a book or onto a website for sharing with others. Students could even partner with younger students to teach about their chosen science concept. These student-created explainer documents are connected to science standards/practices in different ways. Clear communication of findings for different audiences is imperative as students must critically read, evaluate the validity, and explain thoroughly.

Teaching Younger Students/Connecting with Community Members

Both the Children's Guide and the Explainer Creations, as well as the schoolyard mapping and personalized naturalist notebooks, could be used to introduce science concepts to younger students or shared with community members. Students could:

- offer scientific tours using their guides, maps, or notebooks.
- return to the map as seasons change and create overlays or alternate maps for different points of the year. They could partner with a K-2 class that is studying weather or seasons to make the adjustments to the map.
- create scavenger hunts where the maps have to be used to find the answers.
- create videos with tools such as animoto.com or iMovie to share their scientific understandings (i.e., a brain-pop style video to teach others about a topic).
- partner with a city/state park to create a guide for a new trial or observation area or create "kid-friendly" materials of already existing materials.

There are many opportunities that could be designed for teachers to take advantage of to extend the knowledge learned through reading and studying the concepts in the novel. A further consideration is to encourage students to use their maps to advocate for keeping the natural spaces environmentally healthy. We encourage teachers to dream big!

CONCLUSION

The Evolution of Calpurnia Tate provides a point of entry for science students to learn about the nature of scientific discovery—that is, how to make sense of the world that they live in by engaging in scientific practices to help them develop new understandings. Drawing on the before, during, and after reading strategies described in this chapter students will have an opportunity to follow in Callie's footsteps as they have opportunities to develop and refine their use of a variety of science practices in an authentic and meaningful context.

REFERENCES

AdLit. (n.d.). *First lines*. https://www.adlit.org/in-the-classroom/strategies/first-lines.

Burns, L. G., & Harasimowicz, E. (2013). *Citizen scientists: Be a part of scientific discovery from your own backyard*. Henry Holt and Company.

Kelly, J. (2011). *The evolution of Calpurnia Tate*. MacMillan.

Lambert, J. (2022, January 5). *Scientists discover the first true millipede*. Science News Explores. https://www.sciencenewsforstudents.org/article/true-millipede-most-legs-eumillipes-persephone.

Laws, J. M. (2017). *Asking questions, episode 8*. John Muir Laws. https://johnmuirlaws.com/njc-episode-8-asking-questions/.

National Research Council. (2012). *A framework for K-12 science education: Practices, crosscutting concepts, and core ideas*. National Academies Press.

NGSS Lead States. (2013). *Next generation science standards: For states, by states*. Washington, DC: The National Academies Press.

Place-based Education Evaluation Collaborative. (2010). *The benefits of place-based education: A report from the place-based education evaluation collaborative* (2nd ed.). https://promiseofplace.org/sites/default/files/2020-06/PEEC%2C%202010%20summary.pdf.

Ritchhart, R., Church, M., & Morrison, K. (2011). *Making thinking visible: How to promote engagement, understanding, and independence for all learners*. Wiley.

Schanzer, R. (2009). *What Darwin saw: The journey that changes the world*. National Geographic Kids.

Scientific News Explorers. (n.d.). *Explainers*. https://www.sciencenewsforstudents.org/collections/explainers.

Chapter 6

Past and Future Plagues as Windows into the Present

Reading A Death-Struck Year *to Teach about Diseases and Immunity*

David Nurenberg and Ben Lawhorn

The highly localized, highly variegated patchwork system of public schools in the United States ensures widely divergent experiences for students, yet the COVID-19 pandemic created that rare, universally shared point of reference. While geography, race, socioeconomics, and family circumstances certainly shaped students' individual experiences of the pandemic, no American school escaped some length of shutdown, some form of remote learning, and some degree of fear and anxiety, especially fear of the virus' many unknowns. These fears persist even after two years and the ready availability of vaccinations and treatments, underscoring the need for robust education in the science of epidemiology and immunology.

Students always learn science (as well as any other subject!) more readily when teachers can tie it to personal experience rather than present it as an abstract concept (Cavanagh, 2019), yet the personal trauma of COVID may make direct personal connections emotionally fraught for many students. For example, Centers for Disease Control and Prevention (CDC) data estimates that over 100,000 students lost a primary caregiver to COVID, and teachers must adopt particular trauma-sensitive practices that include "exercise[ing] flexibility and empathy" (Crosby, 2020, p. 3) around engaging affected students with any coursework, particularly that which directly relates to the pandemic. Another challenge teachers face when discussing COVID with their students is that "topics like mask-wearing and vaccines are still highly politicized, and teachers say they have to take care to . . . avoid partisan politics or the impression that they're pushing an agenda" (Schwartz, 2021, para. 3).

For all these reasons, Young Adult literature, replete with pandemic tales both historical and fantastical, can provide a necessary remove. Science teachers can use the experiences of fictional characters wrestling with their own pandemics to form a bridge to real science concepts around epidemiology and immunology.

Even before COVID-19, the pandemic subgenre of Young Adult literature was robust (see Hundley & Pendergrass, 2022). From a science education perspective, a major challenge with using many of these texts is that the nature of the illnesses is often so fantastical as to render them practically unusable as tools to study real-world pandemics. Examples include a plague that turns victims into crystalline entities that merge with one another in Kelly McWilliams' *Agnes at the End of the World* (2020), a disease that makes people explode and then compels the survivors to eat the remains in Emily Suvada's *This Mortal Coil* (2017), or the ever-popular zombie outbreaks in any of a dozen other books. To be sure, some elements may still be transferable, and the narratives themselves can be engaging for students, but *A Death-Struck Year* by Lucier (2014) may be a more realistic tool for helping students to personalize and understand concepts such as how diseases spread, how the human immune system functions, disease and vaccination, and managing the "spread" of health misinformation.

A DEATH-STRUCK YEAR BY MAKIIA LUCIER

A Death-Struck Year by Makiia Lucier is a work of historical fiction, whose teen protagonist, Cleo Berry, finds her world shattered by the onset of the Spanish Influenza in 1918. At first sheltered by geography and the privilege of her wealth (although, as an orphan, her life has not been entirely free of tragedy), the crisis that begins only in headlines about far-away people and places eventually comes to roost in Cleo's home of Portland, Oregon. Motivated by her own childhood trauma, Cleo joins the Red Cross and gets thrust firsthand into the horrors of the pandemic—and, typical of many Young Adult novels, it also thrusts her into a hesitant romance with Edmund, the handsome soldier/medical student of her dreams.

PREPARING STUDENTS TO READ
A DEATH-STRUCK YEAR

Assessing Prior Knowledge

Assessing students' prior knowledge before beginning a new learning goal is vital both for adapting instruction to their specific needs and for easing their transition into acquiring new understandings (Dong et al., 2020). The recent COVID-19 pandemic has brought disease outbreaks to the forefront. Although students can draw on their experiences living through this outbreak, teachers may want to assess students' prior knowledge about disease and disease outbreaks beyond COVID-19. Teachers could begin by prompting students to list movies, TV series, books, and video games that have involved disease outbreaks. Students can first write a brief passage or otherwise assemble their thoughts about their favorite example and then respond to prompts such as, *how realistic do you think this outbreak was? Could it actually happen? How did it seem similar to/different from the COVID pandemic?* (a Venn Diagram graphic organizer would work well for that last prompt). They can then, in a "think pair share" model, share their response with a partner or in a small team. By asking the students to contribute to a Google Doc, Jamboard, or just a list on the blackboard, the teacher can aggregate these responses to get a sense of the students' existing knowledge base as well as give the students a chance to start getting familiar and comfortable with discussing the topics involved in a low-stakes way; for that reason, the teacher should emphasize that these activities will not be graded on any kind of content knowledge or accuracy.

While it is tempting to "data dump" with videos and short readings that provide explicit background information on the novel's historical context, the teacher should always root such resources in students' personal experience first. For example, to ground them in the necessary historical background (see table 6.1), before launching into a documentary about the domestic effects of World War I or the economic conditions of the 1910s, such as The Doughboy Foundation's *How World War I changed America* (2022)'s suite of videos and educator tools, students should first be engaged in some sort of role-playing scenario to make such a world feel relatable.

One such scenario could be to ask a student how they would try and treat themselves, or a loved one, who is sick using only the medical capabilities of the times (e.g., no antibiotics, poor understanding of how diseases actually were communicated, etc.). Reading a profile of a nurse from the era, including her firsthand reports of treating influenza victims, will make for a far more engaging introduction to what Cleo will experience than dry facts from a textbook. The National World War I Museum's web-accessible archives include letters and diaries from nurses, and books like Nancy Bristow's *American Pandemic: The Lost Worlds of the 1918 Influenza Epidemic* include many personal accounts of those who treated, and suffered from, the Spanish Flu.

Part of what made the task of such caregivers so difficult was their lack of our contemporary understanding of disease and immunology; the goal of this unit is for students to acquire that vitally important knowledge themselves. Students will be learning the following concepts:

Table 6.1. Background Information Needed

Disease	Characteristics	Available Medications
Bubonic Plague	*Type:* Bacteria *Symptoms:* Fever, headaches, vomiting, swollen lymph nodes *Spread:* Transmitted through flea bites *Mortality Rate:* High	Antibiotics: • Streptomycin (1943) • Gentamicin (1964) • Doxycycline (1967) • Ciprofloxacin (1987) *No vaccine available
Smallpox	*Type:* Virus *Symptoms:* Fever, headaches, body aches, vomiting, rashes, sores on skin *Spread:* Airborne (like COVID-19) *Mortality Rate:* High	Antiviral: • Tecovirimat (2018) *Vaccine developed in 1790s
Cholera	*Type:* Bacteria *Symptoms:* Nausea, dehydration, severe diarrhea, vomiting, pain in abdomen *Spread:* Contaminated food or water *Mortality Rate:* High	Antibiotics: • Doxycycline (1967) • Ciprofloxacin (1987) • Azithromycin (1988) *Vaccine developed in late 1800s
Tuberculosis	*Type:* Bacteria *Symptoms:* Cough, chills, fatigue, night sweats, chest pain, shortness of breath, swollen lymph nodes *Spread:* Airborne (like COVID-19) *Mortality Rate:* High	Antibiotics: • Isoniazid (1951) • Pyrazinamide (1952) • Ethambutol (1961) • Rifampin (1966) *Vaccine developed in 1920s
Spanish Flu	*Type:* Virus *Symptoms:* Muscle aches, cough, chills, fever, fatigue, congestion, sore throat, nausea *Spread:* Airborne (like COVID-19) *Mortality Rate:* High	Antiviral drugs: • Tamiflu (1999) • Relenza (2000) • Rapivab (2014) • Xofluza (2018) *Vaccine developed in 1940s

- Vaccines play a crucial role in preventing the spread of disease and building immunization for deadly diseases. What is herd immunity and how can vaccines help achieve it?
- Modern medicine and understanding of diseases have greatly impacted our ability to prevent the rapid spread of diseases that could potentially result in endemics or pandemics.

- External factors such as working conditions, living proximity to others, or resources can affect the spread of disease. These factors affect different demographics differently and may create injustices.
- The human immune system has three main lines of defense against diseases. Symptoms arise from the body fighting pathogens that pose a threat to our health.
- Diseases can be transmitted in a variety of ways including direct contact, droplets, vectors (such as mosquitos), vehicles (such as food or water), or airborne.
- The virulence of a virus represents the degree of deadliness to its infected host. How virulent a virus is can affect its ability to be transmittable.
- We can protect ourselves from diseases through science-based prevention practices such as wearing masks. How does accessible information or education on a disease play a role in preventing the spread of disease?

A diagnostic activity should be used to assess student understanding of diseases and the scientific terminology associated with epidemiology. Students could engage in a role-play activity in which they devise an "action plan" to prevent their own infection from a disease in the context of their everyday lives. Students are divided into teams of three to four and are given information on a mysterious disease that has been detected. The information they receive will be extremely vague as the disease is relatively unknown. It will include some symptoms witnessed by peers, how fast it seems to spread, and how deadly the disease seems to be. However, no factual information will be given. This scenario would represent how caregivers may have approached new diseases with limited information, like those in the 1910s addressing the initial spread of the Spanish Flu. Students must work collaboratively with the little information given to create an "action plan" to prevent themselves from becoming sick. Teams will then present their "action plans" to the class as the teacher assesses their basic understanding of epidemiology. This activity allows students to present ideas such as vaccines, medications, disease transmission, viruses vs. bacteria, or virulence.

After class-wide discussion, students can revisit their "action plan" with more information on the mysterious disease. Students will now be given factual information on the disease's actual name, documented symptoms, transmission, proven prevention practices, available medications, and virulence. This scenario would represent modern times and the knowledge society has at its fingertips with a contemporary, researched-backed understanding of disease and immunology. Students again will share their "action plans" with one another. As a class, students can discuss how their understanding of the disease changed with more information available and how it impacted their ability to make a viable "action plan." Throughout this diagnostic activity, the instructor can assess a basic understanding of disease and immunology that will be explored in the unit.

WHILE READING *A DEATH-STRUCK YEAR*

The following proposes one potential "road map" in which teachers can help use the novel to teach the science concepts and help students apply the science concepts to make sense of what happens in the plot.

Part I: The Spread of Diseases (pp. 1–50, and revisited later)

Guiding Questions

- How do diseases spread in a population?
- Is the rate constant, and if not, why?
- What factors can influence the rate and reach of diseases' spread?

A Death-Struck Year makes use of suspense to introduce the central disease plot slowly, as a creeping intrusion upon the normal (if not necessarily happy) routines of the protagonist. Cleo hears "shocking stories about Philadelphia and the rest of the East Coast," (p. 19) including overcrowded morgues and such coffin shortages that they were burying people in mass graves with only the clothes on their backs. Louisa's sister had heard of a family who lost a seven-year-old boy. They were so desperate to have him buried in something, *anything*, that they placed him in a twenty-pound macaroni box. A little boy. Buried in a pasta box. As the weeks pass, Cleo's "teachers smiled less and whispered more. Everywhere you looked, students huddled over newspapers . . . a single sneeze was met with a wary glance and a quick backwards step" (p. 35).

As the disease slowly moves from the realm of rumor to Cleo's local reality, it provides a point of reference for the teacher to introduce students to the idea of how diseases spread in a community. In order to model how diseases spread among people, students could make visible how contact among people happens. Using stickers, sticky notes, notebooks, or some other tool, ask students to get out of their seats and "mingle" freely around the room. Whenever they come within arms' length of another student, they can shake hands (or high-five, or put a sticker on the person they met, or just make eye contact) and write the name of the student they've interacted with in their notebook. There should be several rounds of this activity, one to two minutes per round, after which everyone returns to their seats. The students keep track in their notebooks of whom they met up with during each round (repeat meetups are okay).

Secretly, prior to the activity, the teacher has determined which student is Patient Zero in an infection. After all three rounds, everyone returns to their seats, and the teacher announces the identity of the "infected" student. Anyone who wrote that student's name down on their paper during round one, it turns out, became "infected." The class should then look to see if any of *those* students, as well as Patient Zero, were on their list in round two. The pattern continues for the remaining rounds (four to six is a good number, depending on class size), and the teacher can invite students to graph the results on the board and draw conclusions about the rate and patterns of spread. They should see from their experience that epidemics and pandemics start with one person (or with a leap from an animal to a person) and eventually reach a critical mass where the number of infected changes from arithmetic growth (1, 2, 3, 4, 5) to geometric growth (1, 2, 4, 18, 16).

One particularly visual way to run this activity that the authors learned of involved giving students bags with white powder. This powder is flour for most of the students, but for "Patient Zero" it is baking soda. Every time students interact with each other and record their interactions, they mix powders. After multiple rounds, vinegar is added to everyone's bags. If the bag bubbles, it reflects the presence of the baking soda and thus the "infection."

Since not all diseases are spread the same way, the teacher can also make alterations to the activity to reflect the differences between airborne transmission (the students just need to make eye contact) vs. surface contamination (the students touch the same object an infected student touches), and so on. As a higher-order critical thinking and evaluation activity, students could be asked how to make this simulation more realistic. For example, not all individuals in an outbreak have equal propensity to spread disease; some have more contact with others on a daily basis, and so forth. Variations on this activity could involve running it in larger or smaller spaces (a classroom, a gymnasium, or outside on the athletic fields) to see how confined vs. open spaces affect the spread. Toward the end of the unit, this activity can be rerun to illustrate how masking or vaccination rates can interrupt this pattern.

By the time the novel starts showing the spread—Portland sees 200 cases of influenza on October 11th (p. 88), and over 1,500 cases by October 16th (p. 188)—students should be more than able to explain why this is happening (see After Reading Activities for some suggestions).

Part II: Infection and the Immune System

Guiding Questions:

- What external defenses does the human body use to defend itself from pathogens?

- What internal defenses does the human body use to defend itself from pathogens?
- How do these mechanisms connect to symptoms we experience when sick?
- How do we eventually achieve immunity naturally?
- How does vaccination work?

This is a logical order in which to pursue learning the science concepts, but the first three questions are all raised more-or-less simultaneously throughout the rest of the novel, so the teacher may well have to lead the students in some "backtracking" through the book. The final question is more appropriate for the very end of the book (beginning on p. 264) and into post-reading activities.

What External Defenses Does the Human Body Use to Defend Itself from Pathogens?

Part of the terror in *A Death-Struck Year* comes from their incomplete knowledge of infectious vectors and how to protect themselves from disease. Advice Cleo hears ranges from useful—"protect others by sneezing or coughing into handkerchiefs or cloths, which should be boiled or burned"—to irrelevant—"make full use of all available sunshine" (p. 63). At one point, Cleo, with misplaced confidence, assures a concerned Edmund that "I sleep with my windows open. I'll be fine" (p. 138). Today our understanding of disease vectors and our body's defenses are more developed, and the teacher can engage the students in creating a "guide to send back in time" to Cleo and her friends and use this as a frame to introduce the idea of the body's mechanisms of defense against disease through various hands-on activities with the class.

One such activity can involve students attempting to insert objects of varying sizes and compositions (donut munchkins, chocolate chips, and small balls made of copper wire), representing different kinds of pathogens, past various barriers (a wire mesh, a dish of diluted Hydrochloric Acid (HCl)) that represent the body's innate physical and chemical barriers. Students should see and record that the wire mesh (representing skin) stops the munchkins (just as skin can successfully protect the body from certain fungal infections, certain varieties of *staphylococcus, etc.*), but not the chocolate chips, which are small enough to fit through the gaps (such as skin pores), just like human papillomavirus (HPV) and some varieties of Herpes can penetrate skin. The acid (much like stomach acid) will dissolve the munchkins and chocolate chips (given time), but not the copper wire (just as some kinds of *E. coli* and Yersinia survive stomach acid thanks to their tough cell envelopes). Please note that even diluted HCl must be handled with extremely careful attention to lab safety!

This activity can also lead to a discussion of supplementary non-biological barriers, like face masks. Face masks become as much a fixture of Cleo's 1918 world as they are nowadays, so much so that the adjustment to a pandemic-riddled Portland shocks her: "I felt as though I'd stepped into another world entirely. It was the masks" (pp. 50–51). Cleo is no fonder of her mask than many of her real-life twenty-first-century descendants—at one point she bemoans how "I'd been wearing that rotten itchy mask for days" (p. 138). Mask skepticism rears its head in her world as well as ours: Edmund is dismissive of masking, claiming "you might as well try to keep the dust out with chicken wire . . . there are two nurses, one chaplain, and two doctors lying in those cots. All of them wore masks . . . and all of them are sick" (p. 139). On the surface, Edumnd's analogy seems to make sense. The influenza A and B viruses are about 100 nm in diameter while COVID-19 is about 50–110 nm and most woven cloth masks have gaps up to 200 μm in diameter. Nevertheless, 172 observational studies across 16 countries and 6 continents have supported the efficacy, if not the perfection, of mask use as a barrier to aerosol and droplet-based viral transmission (Chu et al., 2020, p. 1973). How can this be?

A spray mister with a Fine, Very Fine, or Extremely Fine spray will produce droplets of less than 200 μm. Students can fill one of these misters with food-colored water and spray it on masks of varying types of cloth, at varying distances, to record how much of the food coloring "makes it through" from one side of the mask to another (a black light might help with this experiment). They will discover that, even at close range, few colored particles make it through. The teacher can then help them understand why these empirical findings seem to contradict the "common sense" nature of the chicken wire analogy, perhaps by showing images of how, at the microscopic level, particle filaments that project between the gaps in the fabric stop many particles (which, unlike mosquitos, cannot maneuver to avoid them). The point is not that no particles penetrate the cloth, but that many are blocked.

What Internal Defenses Does the Human Body Use to Defend Itself from Pathogens?

When pathogens make it past these barriers, the body gears up for a fight. The presence of World War I in *A Death-Struck Year*—reports from the front, the presence of soldiers (e.g., pp. 133–135), the war stories that Edmund tells Cleo (pp. 169–170)—can provide a context for some martial analogies. For example, white blood cells, or leukocytes, can be seen as a "first line of defense," like "first responders" to an invasion, holding the line until better-equipped forces like phagocytes can mobilize. B lymphocytes are like long-range artillery on the battlefield, attacking pathogens outside the cells, while helper T cells are more like the national guard, attacking pathogens that have gained entry to the cell's "home front." Cytotoxic cells are responsible for "executing traitors," cells that have already been compromised and infected (and the mechanisms by which they do so could be compared to the chemical warfare for which World War I is infamous).

Of course, war does not provide the only possible analogy. Students could be encouraged to make their own, new analogies as part of this process, or certainly as a part of final assessment. For example, students could be asked to create an analogy for the different lines of defense in the immune system and the defensive positions in the game of football. Students could be provided with two Term, Information, Picture (TIP) charts: one for scientific terms of the immune system and another for the different defensive positions in football. Students would then work collaboratively to create an analogy that shows the role of each immune system component (or defensive position in football) in preventing infection from a pathogen (or a touchdown from the offensive team). For example, a defensive line in football can be seen as the first line of defense in the immune system (skin and mucous membranes) preventing the offensive team (a disease) from getting over the line of scrimmage (into the body) or scoring a touchdown on the defense (infecting the body). Linebackers represent the second line of defense in the immune system (white blood cells: phagocytes and natural killer cells) preventing the offensive team from gaining additional yards or a touchdown after making it past the defensive line (first line of defense in the immune system). Specialized positions such as safeties and cornerbacks (lymphocytes and antibodies) act as a last resort (the third line of defense in the immune system) recognizing specific dangers (pathogens) such as wide receivers or other passing threats that may result in a touchdown (infection). This analogy can be extended off the football field, as gaining the knowledge of another team's plays (acquired immunity) can be achieved through watching game film and learning from game experiences (previous infections). Furthermore, practicing against a team's plays in a non-game scenario parallels the low-stakes development of immunity that can be achieved through a vaccine. This activity can also be applied to a wide range of student interests outside of football, such as other sports, dance, food, and so on. Students could be given a menu of analogy options to choose from. Original analogies are encouraged, but enough scaffolded analogies would be given to create equitable opportunity for student involvement. Assessment is not limited to writing. Students can be assessed verbally or through art in explaining their analogy.

How Do These Mechanisms Connect to Symptoms We Experience When Sick?

A Death-Struck Year is replete with ghastly depictions of the effects of influenza infection, such as a section from pp. 76–79 which include scenes of a man who "hacked up blood onto a towel while a nurse gripped his shoulders" and patients who "had a peculiar bluish cast to their skin . . . others were so dark they appeared almost black" due to "cyanosis . . . it means his lungs are failing, and he doesn't have enough oxygen in his blood. It can turn a person's skin blue or purpose. Sometimes even black" (p. 77). In one scene, a woman

> struggle[s] onto her elbows. The noise she made was terrible. Wild and rough at the same time, like a cat grappling with a massive hairball. An empty bucket lay on the floor. I dived for it, thrusting it forward just as she leaned over the side and retched. (p. 79)

These passages present opportunities for students to learn how many of these symptoms are actually the result of immune response. For example, the hypothalamus sets the body temperature upward, causing fever, in an attempt to kill off the invading pathogens, and cyanosis results from the body flooding the lungs with fluid in an attempt to create a productive cough to expel the virus. In perhaps the most dramatic of such scenes in the novel, Cleo's friend Kate suffers from a sudden massive nosebleed (pp. 227–228); even this symptom is now understood to be the result of a "cytokine storm," where the immune system gets locked into a cycle of sending more and more cells to fight the infection, causing an overreaction. Students could be engaged with the question of who might be most vulnerable to such "overreactions"—ironically, it is those with the most robust immune systems, including young people like Kate. This realization could hit particularly close to home with adolescents!

How Do We Eventually Achieve Immunity?

It is fortunate for us that our immune system has adaptive powers as well—adaptive powers which, in the end, wind up being Cleo's salvation in *A Death-Struck Year*, as she survives her own infection (pp. 264–268). As a means of exploring this concept, teachers can ask students to think about a time when they saw something as a "threat." As a follow up, the teacher should then ask why they thought this was a threat. What were the criteria they used to determine whether or not something was a threat? Visual examples could work well here: students could be shown an image of a happy, tail wagging Corgi dog beside a growling, snarling German shepherd, or a juicy apple vs. a rotten moldy one, and describe how they came to the determination that one of these objects was "safe" vs. "unsafe?" The teacher can then help students make the leap to how the immune system undergoes a similar process: much like you or the characters in the books, your body relies on prior experiences or learned information to determine what is and is not a "threat."

An activity could involve students, alone or with partners/teammates, making a diagram to track this concept across their own lives, the experiences of characters from the novel, and of characters from other books or movies they enjoy (see table 6.2).

A table is not the only, or even the best, form this product could take: creative visualizations should be encouraged with plenty of supplementary artwork or images from the internet to make the concepts more real. Presenting and explaining these products to fellow students can help cement the concept, both for the presenters and for those who listen (and are held accountable for active listening and construction of new knowledge by writing a reflection afterward).

Previous encounters with harmful antigens are used to identify, prepare for, and fight threats to the body's health. After your immune system has encountered an antigen, it can better identify and fight that threat with already accessible antibodies in the future (i.e., naturally acquired immunity).

Table 6.2. Sample Diagram (created by authors)

	Stimulus	What Made This Stimulus Recognizable as a Threat?	How Did That Recognition Affect Behavior?
Your own life	Sara, the bully from my third-grade class	Her sneer, or when she took my diary without asking and read a page to her friends while laughing and speaking in a mocking tone	I always sat as far away from her as I could on the bus or in the cafeteria
From *A Death-Struck Year*	The balding man in the red sweater Cleo meets at the library on pp. 178–180	He gives her "a sour look" and "pluck[s] the book from [her] hands," he yells at her that the book "is not for public use" (p. 178) and then purposely sneezes in her face, apologizing "without sounding sorry at all" (p. 179)	She gives him "the dirtiest look [she] could manage" (p. 179) and her friend "scowls" at him and scolds him, her voice "loud and indignant"
From another book or movie of your choice	The Dementors from *Harry Potter*	Harry feels cold and hopeless whenever they approach him	He learns how to cast *Expecto Patronum* to create a Patronus to shoo them away

How Does Vaccination Work?

For diseases like viruses that our immune system *can* adapt to, the next logical question is, *do we have to undergo and survive an actual infection, like Cleo does?* (pp. 264–268). *Or, can we skip that dangerous step and "program" the immune system to recognize harmful antigens ahead of time?* The answer is yes—not all immunity has to be acquired naturally. The "encounter" needed to "learn" an antigen can be artificially replaced with information in the form of vaccines. Students can return to their "adaptive immunity" diagram, this time with encouragement to think about a danger they have learned to avoid not through firsthand experience, but through reading or watching a video about it (the teacher can even show a brief airplane or fire safety video—the students now know how to protect themselves, even this event hasn't happened to them directly).

This can also be a time to return to the students' analogy assignment from Question #3. Going with the war analogy, a viral vector vaccine could be thought of as bringing in a captured enemy soldier whose weapons have been removed, so that he can be more safely fought, and after the fight, his tactics and "uniform" (a characteristic protein) can now be recognized by your body's soldiers.

The COVID-19 vaccine may be prominent in students' minds; an mRNA vaccine like this one is more like learning from one of those old wartime "this is the enemy" propaganda films (which the teacher can show examples of), or perhaps intercepted plans about the nature of the enemy's forces; the key is that no direct contact with said enemy is needed now, only the transmission of information. This information is then used to reproduce models of defining features of the enemy, signature weapons and structures, and so forth, so the allied forces can learn how to be prepared for the actual fight when that day comes.

The world's first vaccine was invented by Edward Jenner in the 1790s to treat smallpox, but scientists were not able to create an effective influenza vaccine until the virus was isolated in the early 1930s. Attempts at influenza vaccines were tried as early as 1918, based mainly on pneumonia, and one of these appears in *A Death-Struck Year*, described as "an untested vaccine that was of no use to those already dying. It was not the miracle drug we had hoped for. Far from it" (p. 226). Cleo is concerned that "it's barely been tested" and worries that it will give her rickets or "milk leg" (p. 227). In keeping with history, it is ineffective, as both Cleo and her friend Kate still contract the flu despite their inoculation.

AFTER READING *A DEATH-STRUCK YEAR*

So much of the terror from the novel comes from Cleo's ignorance of the science behind what is happening all around her, and eventually to her, an ignorance that the best scientific minds of her day were unable to dispel. By the end of this unit, the students no longer share that ignorance.

One form of final assessment could ask students to write a "letter to Cleo" to send back in time, explaining the science concepts they have learned to her and giving a "how and why" for the phenomena she has observed—the idea of letter-writing is a familiar one in the novel, as Cleo regularly corresponds with her distant brother Jack. If the students (or the teacher) prefer to empower Cleo a little more, the students can write from her perspective as an elderly immunologist in more recent times, now looking back on that previous epoch of her life.

Another "letter to Cleo" idea could focus particularly on assessing students' knowledge of vaccines: since the novel ends with Cleo taking a rather dim view of their value, students could explain how today's tests are better and more effective than the one she experienced. This is a particularly important activity not only as a check on student understanding from the unit but also because the amount of vaccine misinformation and fear-mongering proliferating on the internet can create the misperception that vaccines' safety and reliability haven't in fact improved much since what we see in the novel.

The teacher might wish to integrate some critical media skills with this activity: one resource for doing this is the National Network for Immunization Information (NNii)'s guidelines for reading online resources about vaccines, such as directing students to the "About Us" pages of vaccine information sites to learn who is paying for a given site, seeking for the original sources of any information posted there, and identifying who (if anyone) with research credentials like an MD, DO, or PhD has reviewed this information.

A more comprehensive test of student mastery of the science concepts from the unit is not only their ability to identify and explain them in terms of how they operate in *A Death-Struck Year*, but also in an entirely different book or movie of either the teacher's or (preferably) the student's choosing, leading to a Venn Diagram, compare-and-contrast essay, or similar product.

EXTENSION ACTIVITIES

Vaccines and the Spread of Disease

Students might be interested in speculating how an effective vaccine might have changed the conditions that Cleo and the other characters experienced in the novel. Since *A Death-Struck Year* ends before depicting the effects of vaccination on the population, a natural subject for extension activities could be to explore those effects. In exploring the effects of vaccination on a population, the teacher can return to the earlier "How Diseases Spread" activity. The first two rounds proceed as before, but this time, Round 3 will include some inoculated students, about 20 percent of the class, each wearing an identifying token like a wristband or sticker. The class will tally the number of inoculation tokens along with how many students became infected—if a vaccinated student gets an entry in the infection notebook, it doesn't count (the teacher may wish to clarify that vaccinated individuals still do get the virus in real life, but their immune systems keep the virus from making them sick).

In each subsequent round, another 20 percent of the class gets vaccinated (so, Round 4 will have a total vaccinated population of 40 percent, Round 5 will have 60 percent, etc.). Students should continue tallying both vaccination rates and infection rates, creating bar graphs of each round to show the accelerating trend.

The teacher should then ask the students, in pairs or in teams, to analyze this trend and come to some conclusions about the effect of vaccination on infection rates. The concept of herd immunity, or vaccinating a sufficient percentage of a population to reduce the likelihood of infection, can be introduced at the end of the activity. Students may be asked to develop their own definition of herd immunity based on their role-play experience and identify when it was achieved in the activity after learning the concept. They should be able to figure out that infection still spreads quickly when only a small number of people are vaccinated, but somewhere between 60 percent and 80 percent marks a threshold beyond which the virus cannot rapidly spread. Additionally, students should be aware that although individuals who are not vaccinated are theoretically "protected" by those vaccinated in the population, non-vaccinated individuals are still highly susceptible to disease if they encounter a pathogen. Herd immunity decreases the likelihood of encountering the disease if you are not vaccinated, not the likelihood of being infected if encountered. This key concept will be explored when students reflect on and analyze the vaccination activity.

Exploring the Human Element

Both vaccines, and any kind of treatment and care for sick people, come as the result of human effort, compassion, and sacrifice. Part of what *A Death-Struck Year* does, like any work of literature, is allow us to see and empathize with those human beings involved, including the victims of disease who are more than just statistics and points on a chart. Students could be encouraged to create "memorials" and extended backstories for the many victims of the plagues in the novel, the major characters, and, perhaps even more importantly, minor characters who appear for only a few moments. Students could begin by finding textual evidence for the facts that they do know, and then exercise their creativity in embellishing these characters' lives. The teacher can require a certain number of additional facts to be created, personal anecdotes of their childhood, favorite activities or foods, and so on. The memorials themselves could take many forms: obituaries, collages, statues, and so forth. Artwork, including a certain number of student-created "photos" from a significant moments in the character's lives (perhaps one textually-tied and one student-created) could be included. Students might even dress up in period clothing and take on the role of re-enacting the character in a presentation or short scene!

First Responders

Students could also create a work that honors first responders or other medical personnel, by name or in general. They could engage in interviews or other research to find out what motivated that person to go into the helping professions (great opportunity for a guest speaker like an experienced Emergency Medical Technician (EMT) or nurse), comparing and contrasting that story with Cleo's motivation for joining the Red Cross in *A Death-Struck Year*, which stems from her own childhood trauma of being all alone after her parents were killed in an automobile accident:

> "No one found me. I waited all night. Until my mother . . . until she passed on . . . . I climbed out of the carriage, out of the ravine, and walked until I found help." She concludes, "You wonder why I stay . . . sometimes I wonder too. But I hate to think of a child, of anyone really, lying somewhere sick and scared, waiting for help that does not come." (p. 171)

Although Cleo often doubts herself and her ability to make a difference as one person in the face of the pandemic, other characters remind her,

> I was not meant to *be* anything at all. "I'm not like you, Hannah," she tells a nurse. "I'm just . . . ordinary." Hannah responds sharply, "Cleo, how many of your schoolmates do you see here? Wearing that armband?" (p. 203)

Similarly, Logan concludes that, in times of crisis, "maybe trying was enough. Or not. But in the end, trying was all you could really ask a person to do" (p. 230).

Education Campaign

The National Center for Child Traumatic Stress specifically encourages teachers in our present pandemic era to help their students "counter expressions of loss . . . with feelings of hope [with instructions like] 'Let's look for the helpers'; 'Would you like to help your classmates collect money for the food bank?'" (Goldman et al., 2020, p. 8). Students could start a disease prevention and/or vaccination education campaign in their neighborhood, putting the science content they learned in this unit to use, along with the inspiration from characters like Cleo, Edmund, and Logan. They could create public service announcements for local television and radio, or share what they have learned about vaccination's benefits at an open-to-the-public information fair. They might even partner with local health providers to set up a free vaccination clinic on school or community grounds and encourage unvaccinated friends and relatives to attend. Not only do these sorts of projects allow students to directly apply learned concepts to the real world (and their teachers to assess that learning in a real-world context), but they also can help students transition from fear and helplessness to a sense of empowerment and purpose.

What Makes Science Research Rigorous?

It is a troubling reality that, in the current political climate, teachers may be pressured to create false equivalencies between well-researched, peer-reviewed studies of vaccines and masking with inaccurate and conspiracy-minded sources. This can be an opportunity for an extension activity that helps students learn about peer review and what makes for a rigorous scientific study, including opportunities to conduct limited-scope studies of their own design on various topics, equipping them with at least some tools to evaluate the validity (or lack thereof) of certain sources on their own without the teacher stepping in as arbiter (and in doing so, facing a potential backlash from parents or administrators).

If the teacher is not facing such pressures, then they may be able to design activities where students can compare and contrast well-known COVID-19 misinformation with relevant passages from the novels. For example, at one point in *A Death-Struck Year*, Cleo encounters a pharmacy selling sham treatments advertised as,

> "Spanish influenza remedies, tried-and-true cures, completely effective." Directly below were displays of mustard tins, quinine jars, onion crates, and baskets filled with Vicks Vapo-Rub. Customers rushed in and out. (p. 51)

Connections could be made to the promotion of ineffective and even dangerous COVID-19 "treatments" like the deworming medication Ivermectin, and an exploration of the psychology behind how and why people are more vulnerable to such false claims in times of crisis. The novel also contains examples of deceptive statements from public officials that minimize the risk and offer false reassurances (p. 146 in *A Death-Struck Year*) of the type that were prevalent in the early days of the COVID pandemic as well.

Pandemics as Social and Biological Phenomena

Finally, students could explore how pandemics are social as well as biological phenomena; not all harm that people suffer during such times is dealt with by the disease itself. The very moniker "Spanish Flu" is a misnomer, as the disease did not in fact have its origins in Spain—students can learn about how "promoting an association between foreigners and a particular epidemic can be a rhetorical strategy for either promoting fear or, alternatively, imparting a sense of safety to the public" (Hoppe, 2018, p. 1462), as has been the case with names for COVID-19 including the "China Virus" or "Wuhan Flu," and the associated uptick in hate crimes against Asian Americans (Health Affairs Forefront, 2022). Although *A Death-Struck Year* does not focus on anti-Spanish prejudice, it does have several passages—pp. 111, 143, 178—that illustrate the effects of wartime-stoked hatred against German Americans.

It's one thing to understand the science behind the disease and medical advances in treating it, but that picture is incomplete without also learning about how that care is inequitably administered. Absent entirely from the novel is any depiction or acknowledgment of the vast inequities in care for white vs. African-Americans during the 1918 Influenza Epidemic, another phenomenon that has echoes and parallels with the racial disparities in modern America's experience of COVID-19.

Students can learn about the social dimensions of disease through teacher pre-selected resources, independent research, or both (for good starting places, try the American Public Health Association, the American Medical Association's *Journal of Ethics*). Students can engage in writing, discussions, and debates, and even actions like the community "Education Campaigns" described earlier.

CONCLUSION

A Death-Struck Year is just one possible YA text that can help add personal, relatable dimensions to the study of disease, immunity, and vaccination, that could otherwise either be too abstract and technical, or, in our present era, too immediate and frightening. While caution should be taken with reactivating student trauma, many trauma-informed education experts believe that "sharing accurate information and science-based facts about COVID-19 will help diminish students' fears and anxieties around the disease and support their ability to cope with any secondary impacts in their lives" (UNICEF.org, para. 2).

Ultimately, both historical and speculative fictions help us to better understand and contextualize the present day. As science provides us with tools to understand the natural world, literature helps us to make meaning of how we put those tools to use or, alternatively, how we ignore them at our peril.

REFERENCES

Chu, D. K., Akl, E. A., Duda, S., Solo, K., Yaacoub, S., Schünemann, H. J., ... & Reinap, M. (2020). Physical distancing, face masks, and eye protection to prevent person-to-person transmission of SARS-CoV-2 and COVID-19: A systematic review and meta-analysis. *The Lancet, 395*(10242), 1973–1987.

Crosby, L., Shantel, D., Penny, B., & Thomas, M. A. T. (2020). Teaching through collective trauma in the era of COVID-19: Trauma-informed practices for middle level learners. *Middle Grades Review, 6*(2), article 5.

Dong, A., Jong, M. S. Y., & King, R. B. (2020). How does prior knowledge influence learning engagement? The mediating roles of cognitive load and help-seeking. *Frontiers in Psychology, 11*, 591203.

Halladay Goldman, J., Danna, L., Maze, J. W., Pickens, I. B., & Ake, G. S., III. (2020). *Trauma-informed school strategies during COVID-19*. National Center for Child Traumatic Stress. https://www.nctsn.org/sites/default/files/resources/resource-guide/trauma_informed_school_strategies_during_covid-19.pdf.

Hoppe, T. (2018). "Spanish flu": When infectious disease names blur origins and stigmatize those infected. *American Journal of Public Health, 108*(11), 1462–1464.

Hundley, M., & Pendergrass, E. (n.d.). *Pandemic/epidemic*. Young Adult Novels Book Lists. https://my.vanderbilt.edu/yabooklists/topics/pandemic-epidemic/.

Lucier, M. (2014). *A death-struck year*. Houghton Mifflin Harcourt.

Schwartz, S. (2021, November 23). COVID-19 is a science lesson waiting to happen. *Education Week*. https://www.edweek.org/teaching-learning/covid-19-is-a-science-lesson-waiting-to-happen/2021/11.

UNICEF Georgia. (n.d.). *How teachers can talk to children about coronavirus disease (COVID19)*. https://www.unicef.org/georgia/how-teachers-can-talk-children-about-coronavirus-disease-covid-19.

Chapter 7

Reading *Ringside, 1925*
Text Support for Teaching Evolution

Frances A. Hamilton and Dana Skelley

The conflict between science and religion has been a polarizing matter for centuries with accusations of heresy falling on early scientists such as Copernicus and Galileo. More recently in the early twentieth century, the scientific theory of evolution was put on trial in a small town in Tennessee when a high school science teacher was arrested for breaking the Butler Act, which banned the teaching of evolution in all educational settings in the state. Even now, evolution is a controversial topic and can be challenging to teach. However, Belin and Kisida (2014) investigated the relationship between state-level beliefs regarding evolution and student performance and found that if the public perceives evolution positively, it has a positive impact on students' achievement in science. In fact, the state's acceptance of evolution was more important than the quality of the standards. Unfortunately, public support may not always be present.

Being a former middle school science teacher (first author), I found one of the most difficult subjects to teach was evolution. For example, there were two occasions in which notes were sent to school asking which days "evolution" would be taught so parents could keep their children at home. After an explanation that most classes consisted of learning about plate tectonics and other geological changes, the universe, and living things, which all evolve, parents realized that students were studying evolution all year.

Parents were not the only detractors of evolutionary content. Students could also quelch discussion and thought, and I learned over time better ways to teach these evolution-related concepts. During my first year of teaching, students were placed in teams and tasked with putting four pictures of fossils into four different layers of rock strata. This was a preassessment to see if students understood that the deeper we dig, the farther back in time we see. When one team proclaimed that they did not believe in fossils, I realized a different approach would be needed in the future; specifically, I learned exposure to information or manipulating real fossils was better than pre-assessments or team activities that might provide naysayers opportunities to share their biases and beliefs with other team members.

Research has shown improving teachers' comfort levels with teaching evolution benefits teachers and students (Fowler & Meisels, 2010). Therefore, we hope our chapter might provide some guidance for teaching the factual knowledge of evolution with the integration of the novel, *Ringside, 1925* (Bryant, 2008). While teaching science concepts without considering religion at all would be ideal, we must acknowledge that introducing evolution often ignites religious-based controversy. Though we cannot completely ignore religion, we also do not want to influence a student's religious beliefs. *Ringside, 1925* can be a safe text to study in conjunction with evolution because some characters in the book hear the scientific facts and reject them due to their religious beliefs, some hear the scientific facts and accept them while maintaining their current beliefs, but none hear the scientific facts and turn away from their current beliefs.

RINGSIDE, 1925 **BY JEN BRYANT**

Ringside, 1925 is a free-verse, fictional account of the true story of the Scopes Monkey Trial held during the summer of 1925 in Dayton, Tennessee. At that time, Tennessee had recently passed a law prohibiting the teaching of evolution in any educational setting. John Scopes was a high school science teacher who was willingly arrested in order to challenge the law after teaching the concepts of evolution in a Rhea County high school biology class. Local business leaders encouraged Mr. Scopes to agree to be arrested as the American Civil Liberties Union had stated they would defend anyone prosecuted under the law, and these leaders wanted the trial in Dayton in order to monetize the circus atmosphere and hopefully impact the small-town's economy (Adams, 2005). Scopes' counsel included Clarence Darrow, a renowned defense attorney, while William Jennings Bryan, former Secretary of State, three-time presidential candidate, and nationally known fundamentalist Christian, volunteered to represent the prosecution for the state of Tennessee. In the novel, the story of the trial and its impact on the community are told through various Dayton students and townspeople. Primary source quotes are sprinkled throughout the novel to reinforce the historical context of this fictionalized account.

PREPARING STUDENTS TO READ *RINGSIDE, 1925*

Concept Development

Prior to beginning the unit, the class will discuss the complexities of the word evolution. In its most basic form, evolution can be described as change over time. Knowing this, the class should create a concept map on the board while they discuss what science-related topics change over time. This might include the following: living organisms (species have common ancestors, structures, and genes); plate tectonics and landforms (to include geomorphic agents, like running water, glaciers, wind, and waves) which create volcanoes, mountains, valleys, and so on; and the cosmos (to include Earth as seen from space, our solar system, and beyond). After discussing the concept maps, teachers can explain that the class will be learning about living organisms, or biological evolution, as they read *Ringside, 1925*. Background information for the teacher can be found on the Geological Society of America's website, including their position statement on teaching evolution.

Knowledge Ratings and Vocabulary Instruction

Next, teachers can gauge students' understanding of essential vocabulary (for the novel and the science concepts) through an activity called Knowledge Ratings. To create a Knowledge Rating graphic organizer, essential terms should be listed in a column on the left side of a page. Three more columns should be drawn and titled from left to right as "I don't know this word," "I've heard this word but can't explain it," and "I can define and use this word." Students are asked to rate their knowledge of each term by putting a checkmark in the appropriate column.

Data from this pre-assessment can help teachers plan what terms need to be covered for vocabulary instruction. For any words marked with I don't know or I can't explain, a student-friendly definition should be given with students logging them in a vocabulary journal for reference. The terms should be discussed and presented in context during whole-class or small-team instruction. Suggested terms with definitions and page numbers for those from the novel are noted in table 7.1.

Story Impressions

For the story impressions activity, students make predictions on what they think the novel will be about. To begin, teachers should display a list of words that are important or indicative of the novel. For *Ringside, 1925* this could include *trial, Tennessee, small town, out-of-town visitors, summer heat,*

Table 7.1. Suggested Vocabulary

Terms	Page #s	Definitions from Oxford English Dictionary (unless otherwise noted)
Evolution	7, 21	process by which different kinds of living organisms are thought to have developed and diversified from earlier forms during the history of the earth.
Butler Act	11, 12, 14	declared unlawful the teaching of any doctrine denying the divine creation of man as taught by the Bible (*Britannica*).
Boardinghouse	14	a house providing food and lodging for paying guests.
Charles Darwin	15	English naturalist whose scientific theory of evolution by natural selection became the foundation of modern evolutionary studies (*Britannica*).
Finches	15	a seed-eating songbird that typically has a stout bill and colorful plumage.
Galapagos Islands	15	island group of the eastern Pacific Ocean, administratively a province of Ecuador (*Britannica*).
Natural Selection	15	the process whereby organisms better adapted to their environment tend to survive and produce more offspring. The theory of its action was first fully expounded by Charles Darwin and is now believed to be the main process that brings about evolution.
Defense	25	the case presented by or on behalf of the party being accused or sued in a lawsuit.
Prosecution	25	the institution and conducting of legal proceedings against someone in respect of a criminal charge.
Origin of Species	28	Darwin started an "abstract" of Natural Selection, which grew into the book, *On the Origin of Species by Means of Natural Selection, or the Preservation of Favoured Races in the Struggle for Life*.
Convict	65	declare (someone) to be guilty of a criminal offense by the verdict of a jury or the decision of a judge in a court of law.
Adaptations	15, 119	a change or the process of change by which an organism or species becomes better suited to its environment.
Analogous Structures	Not in text	similarity of function and superficial resemblance of structures that have different origins (*Britannica*).
Homologous Structures	Not in text	similarity of the structure, physiology, or development of different species of organisms based upon their descent from a common evolutionary ancestor (*Britannica*).

evolution, and *high school science*. Students could write a two- to three-sentence prediction of what they think the novel is about based on the list of words. These could be shared with partners and then with the whole class.

Hands-on Activity: Using "Beaks" to Pick Up "Food"

Early in the novel Charles Darwin, his theory of evolution, and adaptations are mentioned (p. 15). To lay the foundation to support an understanding of evolution, this brief activity about adaptations should occur prior to reading this section. In this activity, students are provided with various tools to mimic birds' beaks (clothespin, tweezers, chopsticks, toothpick, spoon, tongs, and pliers) as well as a variety of seeds in many shapes and sizes. Students begin with a clothespin and try to pick up the smallest seed. They record the number of seeds collected after five attempts at picking up seeds, as well as rate the difficulty level during attempts. They use the clothespin to attempt the next seed, then the next until they have attempted them all. They move to the next tool, the tweezers, and go through the whole process again until they have attempted picking up each seed with each tool, recording their findings along the way. In addition, other "foods" should be incorporated, like gummy worms/fish, to show that some beaks are used to stab softer materials. Next, students record their final findings regarding the level of difficulty picking up various seeds/food. At the culmination of the activity, a direct correlation should be made between the different tools and the beaks of Darwin's finches, pointing out that the birds had different beaks depending on the foods located on the various islands in which they resided, and therefore the birds adapted to their environments. See tables 7.2 and 7.3 for suggested chart formats for data collection and final findings, as well as an example of what a completed chart might look like.

Directions: On the chart below, first record the number of seeds/food collected after five attempts and then rate the difficulty of using the tool using a 1–5 scale, placing the rating in parentheses, where 1 is very easy and 5 is very difficult. Examples are provided.

Table 7.2. Example Data Collection and Final Findings Charts (created by authors)

Seed/Food	Clothespin	Tweezers	Chopsticks	Toothpick	Spoon	Tongs	Pliers
Pea or Okra	8 (1)	5 (2)	3 ()	0 (5)	15 (1)	4 (4)	5 (3)
Pumpkin or Watermelon	4 (4)	7 (2)	2 (4)	3 (3)	6 (2)	7 (1)	2 (4)
Lettuce or Chia	0 (5)	7 (1)	1 (5)	0 (5)	9 (2)	5 (2)	2 (5)
Spinach or Onion	0 (5)	6 (1)	0 (5)	0 (5)	6 (2)	6 (2)	5 (4)

Directions: On the chart below, record your final findings about the difficulty of collecting seeds/food for each tool. Next, answer the question at the bottom of the page to consider how this activity connects to concepts of evolution.

Table 7.3.

Tool	Which Seeds/Food Were the Easiest to Collect?	Which Seeds/Food Were the Most Difficult to Collect?	Notes
Clothespin	Pea/Okra	Lettuce/Chia Spinach/Onion	Small seeds were impossible to grasp with the clothespin.
Tweezers	Lettuce/Chia Spinach/Onion	Pea/Okra	All of the seeds were easy to grasp with the tweezers, but flat seeds were a little more difficult.
Chopsticks	Pea/Okra	Lettuce/Chia Spinach/Onion	None of them were easy, but larger round ones were easier than small or flat ones.
Toothpick	Pumpkin/Watermelon	Lettuce/Chia Spinach/Onion	Could stab some seeds because they were softer. Most were too hard or too small to pierce.
Spoon	Pea/Okra	Pumpkin/Watermelon	The rounder seeds rolled onto the spoon. Could pick up small seeds if they were bunched together.
Tongs	Pumpkin/Watermelon	Pea/Okra	Pea/Okra kept popping out because I was clamping too hard. Small seeds could be picked up if they were bunched together.
Pliers	Pea/Okra	Lettuce/Chia	Small seeds could be picked up if they were bunched together. Flat seeds were hard to keep in my grasp.

Once completed, ask students to discuss how this activity has helped them understand the concepts of evolution. Students should respond that birds' beaks evolved in response to food sources available.

Inventorying Adaptations

Again, to prepare students for understanding the references to evolution and Darwin's ideas of natural selection in the novel (p. 15), this activity expands on the concept of adaptations. Once students explore the importance of features for specific functions with bird beaks, they should recognize the variety of structures and their functions found in many organisms. Here, two approaches are offered, depending on the needs of students, technology and resources available, and time frame desired for study. For the first approach, the class brainstorms as many adaptations as they can, and the teacher lists them on the board. Then teams are created, and each is assigned a different class of organism from the plantae and animalia kingdoms (plants, birds, amphibians, reptiles, fish, and mammals). Students then create posters for different organisms' adaptations. When completed, posters are displayed around the room for a gallery walk.

For the second option, working in small teams, students can brainstorm as many adaptations as they can and create a Jamboard with pictures of several organisms' adaptations. Depending on students' background knowledge, or to minimize repetitiveness, classes of organisms could be assigned. Afterward, students view each team's Jamboards making notations, using the sticky notes feature, about what they notice. Jamboard directions and an example are provided to show what one team's board would look like in figure 7.1 (in this case team 1, the mammals' team). Lastly, a class discussion can occur so students can share what they learned from their classmates' Jamboards. The teacher should create a list of adaptations on the board as students discuss what they noted as a way to document similarities and differences.

WHILE READING *RINGSIDE, 1925*

Court Reporter Character List

To help students keep up with the novel's various characters and their roles, ask students to imagine they are a reporter sent to cover the Scopes trial and are charged with making notes on everyone they meet in the book. Their notes should use the text to describe the characters and what they do and determine their stance on evolution. This character list can be used as a reference while reading when students need a reminder of which character is which. This list also can be used as a study guide for an end of novel test. Table 7.4 contains a key of character descriptions and its format can be given to students as a template to create their own charts. Note there are brief author-created descriptions in the pages before the text begins of the nine characters who narrate portions of the novel. Students should add more information for these main characters and teachers may decide which of the other characters listed in this key they want their students to add to their lists.

Jamboard Directions

Step 1: Each group will be assigned a class of animals or plants (see below).
Step 2: On a piece of paper, students will list as many adaptations as they can think of for their class of organisms.
Step 3: Each group will create a Jamboard slide with pictures of organisms. Try to use pictures that will help indicate the adaptation you want your peers to identify. Your goal is 10 different adaptations.
Step 4: Make sure pictures are numbered. This will help us with the next step as well as during class discussion.
Step 5: Visit each Jamboard slide and, based on the number system below, write on the Jamboard sticky note what your group believes to be the adaptation. Please label each sticky note with your group number. Let's also use the same color on all sticky notes throughout the activity.

Group 1: Mammals group-yellow sticky notes, comment on pictures numbered 1 and 6
Group 2: Insects group-pink sticky notes, comment on pictures numbered 2 and 7
Group 3: Birds group-blue sticky notes, comment on pictures numbered 3 and 8
Group 4: Reptiles/Amphibians group-green sticky notes, comment on pictures numbered 4 and 9
Group 5: Angiosperms (flowering plants) group-orange sticky notes, comment on pictures numbered 5 and 10

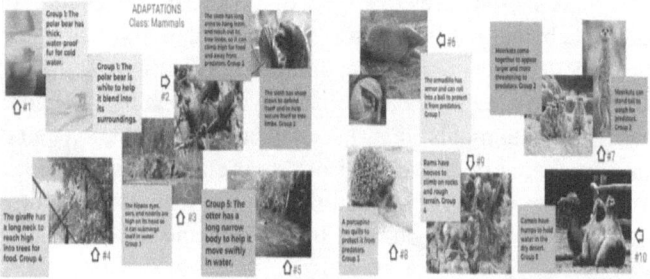

Figure 7.1. Directions and Sample Jamboard (created by authors).

Guided Reading Discussion Questions

The following questions can be utilized during reading for guided discussion, giving teachers an opportunity to help students make connections between the characters' various responses to the teaching of evolution and between what they are learning about adaptations and how that fits in with the general theory of evolution noted in the novel. For students needing less scaffolding, questions can be given to small teams for discussion that is later brought back to the whole class, and with any method, students can be asked to write down their thoughts either before or after discussion, and these are taken up for an informal assessment of reading comprehension.

Part One

- Why is Mr. White looking for Mr. Scopes? (p. 3) Use pages 6–9 to summarize the reason there is going to be a trial.
- What does Willy tell us about the town of Dayton, Tennessee, on pages 4–6? Re-read Willy's thoughts on page 18 and Tillie's observations on the

Reading Ringside, 1925 123

Table 7.4. Character Descriptions

Character Name(s)	Character Information
Peter Sykes	A student at Rhea County High and friend to Jimmy Lee Davis. He works at Robinson's Drugstore. He is interested in learning more about evolution. pp. 3, 11
Jimmy Lee Davis	A student at Rhea County High and friend to Peter. He also works at Robinson's Drugstore. He feels that evolution goes against his religious beliefs. p. 6
Willy Amos	A twelve-year-old African American boy who does not attend school but is educating himself. He works with his dad, a handyman, completing odd jobs and tasks during the trial. He is interested in learning more about evolution. p. 4
MaryBeth Dodd	A student at Rhea County High and lives with her widowed father. She and Peter like each other. She works with her Aunt Tillie at the Mansion, a boarding house, and hopes to further her education after graduating high school. pp. 12 and 13
Tillie Stackhouse	A resident who manages the Mansion. She is the cousin to MaryBeth's father, Frank. She is interested in learning more about evolution and is supportive of her niece's efforts to attend college. p. 14
Betty Barker	A resident who is very outspoken about her religious beliefs against evolution. She is a member of the ladies Bible study group. p. 19
Ernest McManus	A Methodist minister visiting Dayton specifically to see the trial and does not believe the tenants of evolution conflict with his Christian beliefs. p. 33
Constable Fraybel	A local law enforcement agent who is stationed at the trial. p. 23
Paul Lebrun	A young reporter from *St. Louis Post-Dispatch*. p. 46
Minnie Bly and David Amos	Other residents of Dayton briefly mentioned include Minnie Bly, known as the crazy mountain lady p. 39, and David Amos, father to Willy. p. 38
Mr. Walter White	The superintendent of the Rhea County school system. He was one of the men who asked Mr. Scopes if he would mind being arrested for teaching evolution. p. 3
Mr. Fred Robinson	The owner of Robinson's Drugstore. He was one of the men who asked Mr. Scopes if he would mind being arrested for teaching evolution. pp. 3, 7
John Scopes	The high school teacher who taught concepts of evolution in the Rhea County high school. p. 3
Clarence Darrow	A lawyer for the defense working to revoke the Butler Act because he believes the science of evolution should be taught in schools. pp. 29, 43
William Jennings Bryan	A lawyer for the prosecution working to uphold the Butler Act. He is a national celebrity and does not believe evolution should be taught. p. 29
Judge Raulston	The local judge who presides over the trial. His rulings imply he is partial to the exclusion of evolution in the classroom. p. 75

top of page 28. Why are these passages important to understanding why the trial took place in Dayton?
- Concepts of evolution are mentioned on pages 7, 15, and 21–22. Describe how the concept of adaptations we have studied so far fit with these concepts.

Part Two

- How do you think Mr. Robinson feels about evolution and the trial being in Dayton based on evidence from the text (pp. 29–32)?
- From pages 39 and 40, contrast the actions of Betty and Minnie toward Willy. Why might some people be surprised by how each acted?
- On pages 43–45, MaryBeth Dodd shares various descriptions of Clarence Darrow. How do Daddy's (Frank), Tillie's, and the newspaper's descriptions of Mr. Darrow differ and what are the possible reasons for those differences?
- In the section from pages 48–51, Jimmy explains his feelings and reasons for why he does not want to see Mr. Darrow at the train station. Using a quick write strategy, predict what you think will happen between Pete and Jimmy as the trial gets underway.

Part Three

- Information shared through Ernest McManus on page 59 outlines why William Jennings Bryan was a national celebrity. Create a Venn Diagram to compare and contrast Bryan to Darrow using the Bryan information from page 59 and the Darrow information from pages 43–45 discussed above.
- Read Jimmy's account of his conversation with Pete on pages 67–69. This was a turning point in the book that showed the divide occurring between the two friends. In an effort to consider how religion and science are coming between Jimmy and Pete, rewrite the dialogue so that each is respectful of the other's beliefs. Turn and talk with a partner before sharing in teams or with the whole class.

Part Four

- After reading pages 90–91, summarize why Mr. Neal wanted the judge to quash the charges against Mr. Scopes.
- Cite text on pages 92–93 to explain the three reasons Mr. Scopes' defense lawyers gave for the Butler Act being unconstitutional and the three reasons the prosecution lawyers gave for why Mr. Scopes should be charged with breaking the law.

Part Five

- Drawing on pages 114–115, discuss if Mr. White is as guilty as Mr. Scopes based on Constable Fraybel's three points. Use textual evidence to support your reasoning.

Part Six

- Read pages 147–149 and compare the way *The Great Gatsby* is perceived by Betty Barker and Tillie's sister, Lila.

Part Seven

- After reading pages 175–186, discuss why Mr. Darrow called William Jennings Bryan to the stand. How did it help or hurt Mr. Scopes' case?

Part Eight

- Describe how MaryBeth's father's attitude toward her continuing with school change throughout the book (see also pages 213–215).
- Why do you think he changed his mind?

Epilogue

- Note up to three things that surprised you about how events unfolded after the trial. Discuss what you thought would happen and why you were surprised.

Identifying Reliable Sources

An important aspect of the scientific argument is using high-quality evidence to support claims. In this activity, students are presented with one or more quotes from *Ringside, 1925*. Next, the teacher asks questions regarding a claim related to the quote(s). Then, students determine whether the quotes are good sources of evidence for the claim and prepare themselves to explain their thinking. Students will share their thoughts with a partner prior to whole-class discussion. Table 7.5 contains several quotes from the text and questions for consideration.

AFTER READING *RINGSIDE, 1925*

Exploring Analogous and Homologous Structures

After reviewing the novel excerpts for the reliable sources activity, remind students that those descriptions each illustrate that things change over time.

Table 7.5. Quotes and Questions for Identifying Reliable Sources

Character	Page and Quotes	Questions	Responses
Peter	*Now my dream is to ride in an airplane through the Rocky Mountains . . . to see for myself the layers of sediment built up over centuries* (p. 11) *. . . and those places where the earth's crust broke open, pushed up miles of new peaks.* (p. 11)	Discuss why these quotes may or may not be good evidence for the claim "The earth changes over time"?	Students might say yes because the character talks about things that change over time, but Peter is a character (high school age) in a book with no expertise on this topic.
Tillie	*. . . trip on the HMS Beagle to the Galapagos Islands, where he [Darwin] spent several years observing the different animal species and where he found the evidence he needed to support his idea of "natural selection."* (p. 15) *I'd read an article about it [natural selection] in my National Geographic, where they had pictures of lizards larger than me (imagine!) and finches with all different kinds of beaks, depending on which island they lived on and what they had to eat. . . .* (p. 15)	Discuss why or why not these quotes provide good evidence for the claim "There is support for the idea of natural selection?" What are some other reliable sources?	Students might respond yes because the information comes from National Geographic. In addition, students have completed the bird beaks as tools activity, which provided insight into the idea of natural selection. Still, during discussion, it should be pointed out that Tillie is presenting her interpretation of what she read. Affirm that National Geographic is a reliable source if we are using it first-hand. Discuss other reliable sources in addition to National Geographic. Scholarly databases, like EBSCO; websites, like science.gov; peer-reviewed journal articles found online, like on Google Scholar; news sources, like *The New York Times* or *Washington Post*; and professional standards organizations, like The American Psychological Association (APA). Consider: currency, relevance, authority, accuracy, and purpose of the source.

Reading *Ringside*, 1925   127

Jimmy Lee	He [Mr. Scopes] assigned us the chapter in our textbook on evolution, & he reviewed some of the main ideas.... like how the earth was once too hot for any life-so hot you could fry and egg almost anywhere on the ground. But then it cooled & there were some teeny-tiny single-celled animals...; & then later, sea creatures... & then slowly over years & years, some of them grew legs...& they came up on land & changed over millions more years, into reptiles & mammals. (pp. 21, 119 as he explains what happens at the trial)	Discuss whether or not this quote is a reliable source for the claim that evolution exists. Describe the claims made in this quote. What are possible reliable sources to check these claims?	Students might think this is reliable evidence because the information originated from a textbook. However, this is a student's account of what was in the textbook, so it is not a reliable source. Discuss that textbooks are not always a reliable source and possibly share examples of misinformation found in textbooks. Extension: make a list of all the claims made in this quote. Then, discuss possible sources that could be used as evidence for the claims.
Tillie	He [Dr. Metcalf, zoologist] said the term evolution means "the whole series of physical changes which have taken place during hundreds of millions of years" which resulted in "the change of an organism from one character into a different character [in terms of] its structure, or its behavior, or its function." (p. 132)	Discuss if this quote is a reliable source for the claim that evolution exists.	Students might be torn here because Tillie is quoting an expert. However, they might consider that this is a historical fiction text, which could have them wondering whether this part is fiction. Discussion should lead to the fact that the author did conduct a lot of research from first-hand sources for this book and that, in researching, we can see that Dr. [Maynard M.] Metcalf was an actual zoologist called as an expert witness in the trial. Emphasize that students should go to first-hand sources as evidence for claims. More information about the testimony, including what is in this quote, can be found in the description of Maynard Metcalf on the famous Trial link located on the University of Missouri-Kansas City Law School site.

To continue exploring this idea, students will look more closely at the features of different species: some which differ in structure but have similar functions, or analogous structures, as well as those with structures having similar features, but possibly dissimilar functions, or homologous structures. Analogous structures include structures such as the wings of a moth, fly, and bird which developed independently as adaptations for the common function of flying. For this activity, have students look at photos to explore analogous structures. There is one online that compares the following: ant leg to horse leg; fish fin to dolphin flipper, and cicada wing to a chicken wing. Students can record their observations and findings. Then, students can compare homologous structures, such as human, horse, bat, and bird appendages because they have similar features indicating the same developmental origin. There is one online that compares the humerus, radius, ulna, carpals, metacarpals, and phalanges of a turtle, dolphin, horse, human, chicken, and fruit bat. Students will record their findings and discuss in small teams the similarities and differences in appendages and how the different appendages help organisms survive in their environments.

Important Word

Using a variation of the strategy, Most Important Word, teachers should give students the following list of words: *natural selection, law, evolution,* and *adaptations*. Students should choose one of these to connect to various components of the novel. Next, they can create a graphic organizer by placing their chosen word at the top of a piece of paper and describe in no more than three sentences in the given sections how that word impacts the novel's characters, conflict, plot, and setting. Once their graphic organizer is complete, students can be assigned to teams to discuss their word choice and how they felt the word is reflected in the story. An example of a graphic organizer is offered in table 7.6.

Unsent Letters

After completing the novel and science unit, students will be asked to write a letter to a friend. The letter should contain (1) a description of what was learned about adaptations and (2) a discussion of how the new knowledge relates to the Scopes trial and those involved in the novel. Proper letter format should be used, and content should be explained as though the friend did not attend the trial and has never heard of adaptations. Encourage the use of the essential vocabulary first introduced in the Knowledge Ratings by posting the list of words used in the before-reading activity. Students should refer to those as they draft their letters and utilize at least three terms specific to the trial and three terms specific to evolution.

Table 7.6. Example Graphic Organizer (created by authors)

Important Word: Evolution	
Characters Evolution impacts the characters in two ways. It is the concept that is taught that everyone has to decide their beliefs about, and some of the characters go through an evolution as they learn to live with each other's differing beliefs like Jimmy and Pete.	**Plot** Evolution is the main topic of the plot. There would be no story since the trial is occurring because a teacher in their town taught the concept of evolution.
Conflict The conflict in the book is how some people think having evolution taught in schools is okay and others do not.	**Setting** The concept of evolution plays a part in the setting because leaders of the town make a plan to have Mr. Scopes arrested (with his permission) thinking the trial will bring much business to their shops. The trial happened in Dayton because the town leaders wanted it to happen there.

EXTENSION ACTIVITIES

Creating Organisms with Adaptations for Surviving Specific Environments

After a whole-class discussion reviewing examples of ways animals adapt to their environments (camel hump holds water to survive the desert, brightly colored flowers attract pollinators, etc.) and to show knowledge of adaptations, students can choose an existing biome or describe an entirely new one to then design and build a fictitious organism that can survive in that environment. First, students write out a description of, or sketch, a biome. Then, they sketch an organism, creating adaptations that will allow the organism to survive in the biome. Students list these adaptations in their sketch. Next, students use everyday items (either brought from home or provided by the teacher) to construct their organisms. The organisms should contain a minimum of three adaptations. Student-made creations could include organisms with a sponge body to better absorb and hold water in desert environments, an animal with foam-padded feet for stalking upon prey in the savannah, an organism made completely out of white materials to blend in with the snow in the tundra, or an organism with a waxy body to repel water in the jungle. For example, a student might build a jungle plant that has a waxy body to repel extraneous amounts of water, as well as brightly colored flowers to attract pollinators, and large leaves to absorb more sunlight through a dense canopy. Some materials for construction might include sponges, paper plates, foam pieces, aluminum foil, wax paper, clothespins, popsicle sticks, cotton balls,

Q-tips, white and colored cups, straws, wiggly eyes, paper, tape, staples, and brads. Students should plan, sketch, and discuss ideas with peers prior to constructing their organisms. If students will bring objects from home, guidelines should include the absence of dangerous or hazardous components, like sharp objects. Additional guidelines could include size requirements, especially if constructions will be displayed or transported. Recommended graded components include: feasibility of adaptations for biome; number of adaptations; how well the construction follows the sketch (which can be adjusted during building phase); and how clearly students presented their organism to classmates.

Students can present their created organisms to the class describing the adaptations needed to survive in the environment they selected or created. First, students should describe the environment, including the average temperatures and precipitation, as well as the terrain. This information could be presented in a digital format or through a hand-made poster. Then, they briefly explain the materials used to create their organism and their general functions, pointing them out on their 3D model. Next, have students detail how the structures, or adaptations, of their organism help it survive in the environment. Finally, students should explain how their organism's structures relate to organisms with which we are familiar, if applicable.

An example might be presented like this: I created an organism to survive in the desert, which has a hot, dry, and sunny climate. It receives less than 10 inches of rain in a year and reaches over 100 degrees Fahrenheit during the day. I used a sponge for the body of my organism, straws for legs, and large wiggly eyes (pointing at each item). I also rolled up a piece of paper to give it a pointy beak. The sponge helps absorb and hold water for long periods of time. It can push its long legs down through the sand to suck up water found deep under the surface. My organism has large eyes to see predators coming from afar since there is nowhere to hide, and the pointed beak allows my organism to poke through the tough skin of a cactus to find more liquids. My use of a sponge can be related to a camel's hump holding water. The legs make me think of plant roots reaching into the ground for water, and the large eyes make me think of an owl's ability to find prey in its surroundings, only I am using them to watch out for predators.

Annotated List for Further Reading

This further reading extension activity can be used as an independent study. The titles suggested offer fictionalized and non-fiction options for a deeper dive into the science and novel themes. Students can independently read a title of their choice and then choose a presentation option for sharing what they learned with the class. Presentations could include designing a comic strip summarizing content, creating a clothespin timeline of important events, recording a song summarizing content, filming a movie trailer summarizing content, and so on.

- *Monkey Town: The Summer of the Scopes Trial* by Ronald Kidd (288 pages)
  In 1925, fifteen-year-old Francis Robinson is caught up in the "trial of the century" when her father helps get John Scopes (her secret crush) arrested for teaching evolution. The two-week-trial upends Francis' inner and outer worlds as famous thinkers descend on their sleepy town to debate the origin of humans.
- *Summer for the Gods: The Scopes Trial and America's Continuing Debate Over Science and Religion* by Edward J. Larson (336 pages)
  Larson's Pulitzer Prize winning novel seeks to remove myth and legend from the 1925 Scopes trial held in Dayton, Tennessee, and allows readers to consider how science, religion, and public education can coexist. This text is most appropriate for secondary school readers, but excerpts could be utilized for younger readers to flesh out more information on the trial and the repercussions of the trial still with us today.
- *What Darwin Saw: The Journey that Changed the World* by Rosalyn Schanzer (48 pages)
  This non-fiction picture book tells of Charles Darwin's historic trip on the *HMS Beagle* where he began developing his theory of evolution. Using direct quotes from Darwin's journals, this text may be designed for younger readers, but the rich content can support secondary school students' comprehension of Darwin's journey and his revolutionary discoveries.

CONCLUSION

Teaching evolutionary concepts can be a tricky task in secondary classrooms. Ideally, religion would not be discussed in those settings, but with parent concerns, it can sometimes be unavoidable. This chapter suggests pairing a unit on evolutionary adaptations with the Young Adult novel, *Ringside, 1925*. This text explores the events surrounding the Dayton, Tennessee Scopes trial and offers varying perspectives of several characters.

The science concepts and activities presented are designed to explore evolutionary adaptations with students and include the practice of key science skills through implementing hands-on activities; recording observations and results; analyzing data; discussing results in small teams; organizing information onto posters or Jamboards; viewing the work of others during a gallery walk; analyzing and determining reliable sources; and planning, creating, and presenting models and environmental information. It is our goal with this chapter to help teachers present evolution in an engaging way through the use of quality teaching techniques and also in a non-threatening way by letting students see through exploration of the *Ringside, 1925* text that those characters did not let science impact their religious beliefs.

REFERENCES

Adams, N. (2005, July 5). Timeline: Remembering the scopes monkey trial. *NPR*. https://www.npr.org/2005/07/05/4723956/timeline-remembering-the-scopes-monkey-trial.

Belin, C. M., & Kisida, B. (2014). Science standards, science achievement, and attitudes about evolution. *Educational Policy*, *29*(7), 1053–1075.

Bryant, J. (2008). *Ringside, 1925*. Yearling.

Fowler, R. S., & Meisels, G. G. (2010). Florida teachers' attitudes about teaching evolution. *The American Biology Teacher*, *72*(2), 96–99.

Simpson, J. A., Weiner, E. S. C., & Oxford University Press. (1989). *The Oxford English dictionary*. Clarendon Press.

Chapter 8

Studying Genetics and Ethics through Young Adult Literature
How The Gardener *Can Harvest Student Engagement in Biology*

Janine J. Darragh, Ashley S. Boyd, and Kristina L. Podelnyk

Understanding science is the gateway to understanding and navigating our world. Through the lens of experimental design, students learn how to collect evidence, test theories, and communicate conclusions. Unfortunately, often traditional science classrooms leave little space for exploring the ethics of science. While significant time is granted to tackling questions such as *"how can we test this hypothesis?"* very little time is left dedicated to addressing the question, *"should we test this hypothesis?"* Yet, we know in the real world, these two quandaries generally walk hand-in-hand: just because you *can* do something, *should* you? Exploring genetics and inheritance, two topics tackled in every secondary science classroom, offer the perfect opportunity to learn scientific concepts while addressing ethical questions we face today.

Alone, genetics (clarifying relationships about the role of DNA and chromosomes in coding the instructions for characteristic traits passed from parents to offspring) can be a difficult concept for secondary students to grasp. Between Punnett squares, the chemical composition of DNA and mitosis, the practical application of core genetic principles can be lost in the details (and the mountains of vocabulary). Young adult literature in general and *The Gardener* (Boden, 2010) specifically offer a vehicle through which students can be introduced to practical (and ethically dubious) applications of genetic engineering in a fictional, and therefore somewhat removed, space. Rather than jumping into a discussion on the complexities of scientific ethics with regard to current events, students can ease into the concept, first discussing fictional characters and then contemplating both the choices they make and the repercussions of those choices. As such, *The Gardener* (Boden, 2010), on which we focus in this chapter, is a powerful addition to units of scientific study on both genetics and ethics.

THE GARDENER BY S. A. BODEN

The Gardener (Boden, 2010) tells the story of Mason, a high school sophomore, whose mother works at a local care facility run by the company TroDyn, a research institution. Mason's father has been gone most of his life, and he is left with only a video of his father reading a children's story aloud. After experiencing a brutal dog attack at a young age, Mason's face is scarred; he grows into a large size and manifests as a protector, a hero against school bullies.

Soon into the novel, Mason meets a teenage girl at his mother's place of employment who mysteriously responds to the video of his father reading, coming alive, and then retreating to a comatose state when the video is played and stopped. Mason helps the girl escape, and so begins his journey learning about the scientific experiments at TroDyn to create self-sustaining "human" beings who are autotrophs, able to survive without food or water. He also learns of his own parents' involvement in the experimental trials, particularly encountering his mastermind father who is known as "The Gardener." *The Gardener* forces not only reckoning with complicated family decisions but a consideration of larger social issues including bioethics and to what lengths genetic engineering should go to save our population. *The Gardener* is a story about Mason's journey as he comes to terms with his own morals, beliefs, and loss, but it also examines real-world ethical issues, including human experimentation and consent, through the perspective of a teenager grappling with his own identity and future.

PREPARING STUDENTS TO READ *THE GARDENER*

Any pre-reading activity should focus on students' thinking and communicating their thoughts, as ethics is about talking, sharing, and justifying. Through these critical conversations, the teacher sets the stage for both the introduction of the young adult text and also the science unit on genetics and the ethical practices inherent in both.

Introducing Ethics

While students will undoubtedly be familiar with the term "ethics," it is crucial for teachers to begin a study of *The Gardener* with a full consideration of ethics and science and for students to unpack the fluid boundaries therein. First, teachers can begin by asking students to construct their own definition of ethics as a general term and then brainstorm where a person might learn those. Building from this, youth can discuss if these are different from a broader set of social ethics and how conflicts between the two can be resolved.

To acquaint students with the connections between science and ethics, teachers can pose several statements and have students take a stance in a Four-Corners activity, where students determine if they agree, strongly agree, disagree, or strongly disagree with a statement. Statements that teachers could utilize include:

- Funding for scientific research should be determined by a set of experts without the input of the general population.
- Research methods, even if to obtain vital information or make groundbreaking discoveries, should only entail painless procedures for humans and animals.
- Scientific data should only be used for the benefit of humanity.

Teachers can call on students from each corner to explain their response and encourage those on different sides to consider what they hear, allowing them to shift positions if they are so moved. This can help students discern how these issues can be more complicated than they may initially believe and will set them up to explore more pointed examples.

Teachers might go one step further, and, instead of giving students four options, have them take a direct stance on an issue. In this case, teachers might encourage students to dig deeper into a single ethical question, taking a stance and justifying their response. Some example questions connecting to *The Gardener* specifically are:

- Is it ok to use plants in science experiments to discover cures, treatments, and medical breakthroughs?
- Is it ok to test new treatments, medicines, and products on animals?
- Is it ok to use humans in scientific experiments, like testing vaccines, medical treatments, and products?
- Is it ok to include infants and children in scientific experiments?

One of the many strengths of using *The Gardener* in conjunction with a science unit on ethics is that there are no "right" answers. There are questions about consent, and who has the power to grant it. There are also questions about whether the ends justify the means. In the novel, Dr. Emerson shares these same ethical questions that she had as a researcher at TroDyn. When she explains her reasoning for participating in the study on children, Mason asks her, "And experimenting on kids is so moral?" She replies, "For the greater good, I say it was. The parents had the right to decide for them." She continues with,

> to be able to fix a universal problem that's only going to get worse? . . . We were doing something that mattered. It was a sacrifice, but it was the needs of a few against the needs of many. The needs of the many will always be more important. . . . Don't they have to be? (pp. 139–140)

Before students even open the novel, they can discuss these questions, then they can revisit them as and after they read, noting if, how, and why their opinions have changed.

Formative Probe

The Gardener provides an opportunity for students to explore genetics, ethics in science, and ethical experimental practice. In order to further explore these complex topics, teachers might first want to determine what students know about genetics beyond the Punnett Square and making paper chains of DNA—two activities that they most likely encountered in science classes. One way to pre-assess students' understanding is through a formative probe activity—a dialogue scenario whereby first a scientific phenomenon is encountered, then students explore potential explanations by listening to a diverse group of people explaining their understanding of the topic (Keeley, 2021). Finally, students are given space to reflect on the presented perspectives, choose a speaker with whom they agree, and justify their choice. The purpose of this formative assessment is to elicit whether students have preconceived ideas about the scientific concept presented. In this case, the scientific concept should explore how genetic traits are inherited. Scenarios might revolve around inheritance patterns of mammal fur color (National Science Teachers Association [NSTA], 2007), flower petal color, or any other trait that follows traditional dominant/recessive inheritance patterns.

For example, modeling after Keeley's (2021) probes, the science teacher might provide a scenario regarding a group of friends discussing the fur color (white or black) of a litter of kittens. Students can read provided explanations regarding fur color and choose the explanation with which they agree and explain why (see figure 8.1). If teachers are uncomfortable with or do not have time to create their own dialogue scenarios, they might simply turn to Keeley's (2021) *Uncovering Student Ideas in Science: 25 More Formative Assessment Probes*, which offers probes for physical earth, space, and life science assessments. Chapter 17's "Baby Mice" chapter would work especially well in conjunction with a study on inheritance patterns in mammals.

Micah's pet cat had kittens. Five of the kittens were black and two were white. The father cat was black. The mother cat was white. Micah and his friends wondered why the kittens were different colors. These were their ideas:

Janice: Kittens inherit more traits from their father than their mother.
Mason: The kittens got half their traits from their father and half from their mother.
Aleshia: Male traits are stronger than female traits
Fionn: The black kittens are probably male and the white kittens are probably female.
Lester: Parent's traits like fur color don't matter - nature decides what something will look like.
Brandi: Blood type determines what traits babies will have.

Which response do you most agree with and why? Explain your thinking.

Figure 8.1. Sample Kittens Discussion (Modified from Paige Keeley's Uncovering student ideas in science: 25 more formative assessment probes. (2nd ed.). NSTA Press).

The value of using this type of observed dialogue formative assessment probe is in risk reduction. It can be scary to speak out loud in class (especially to share ideas or explanations in science) because students are primed to believe there are only *right* and *wrong* answers. And it can be terrifying to be incorrect, publicly. Dialogue scenarios are a way to invite students into formative scientific conversations in a safe way (Keeley, 2021). Students aren't coming up with their own explanations; they are choosing a response already provided and justifying their choice. Moreover, teachers can use the students' responses as a form of pre-assessment to determine what their students might already know about the topic being explored.

Wondering

As a follow-up pre-reading activity to the formative probe, teachers may introduce current events in the media that have ethical components, be it stem cell research, genetically modified organisms, designer babies, or even going back to the 1997 news story of Dolly the sheep. Students can read or watch a piece from pop culture and discuss how the topic makes them feel as well as identifying what they wonder about the topic. These "wonderings" can serve as future research opportunities or topics to return to throughout the unit of study. In an activity like this, the goal is to make the content applicable and approachable to students- the topics introduced in *The Gardener* are not necessarily just science fiction; similar ethical considerations are happening across the globe on a daily basis and have the potential to affect students' lives directly.

WHILE READING *THE GARDENER*

Alongside reading *The Gardener*, it is important to make space for strategic conversations regarding the choices of the main character in addition to timely learning experiences for content acquisition. There are a number of powerful classroom and laboratory activities that effectively teach dialogue strategies and the fundamental principles of the molecular basis of heredity. What follows are just a few recommendations.

DNA Extraction

To explore the scientific concept of genetics, teachers should have students engage in a variety of laboratory experiences while reading *The Gardener*. A great lab protocol to launch the unit is DNA extraction (Science Buddies, 2013). This activity is simple and exciting and invites students to touch and engage in science, where, more often than not, principles and concepts are traditionally abstract. A myriad of protocols and activities are available through a simple internet search, but one that science teachers seem to find the best and most successful is to extract DNA from a strawberry. Variations include extracting DNA from cheek cells, bananas, and yeast cells. Sometimes, it is valuable for students to extract DNA from multiple sources, so they can compare the results.

DNA STRAWBERRY EXTRACTION LAB

In this lab, students will extract DNA from strawberries that have been blended with water. A portion of the strawberry mixture is then treated with shampoo and salt, mixed for five to ten minutes and then strained through cheesecloth. The filtrate is added to cold alcohol and the DNA from the strawberry solution precipitates (becomes visible).

Prediction: What do you think the DNA will look like? _____

Materials:

20 mL of distilled water	(1) 4 in × 4 in cheesecloth	(1) stirring stick
table salt	(1) plastic pipette	(1) plastic cup
clear shampoo	(1) test tube of isopropyl alcohol	container of strawberries
50 mL beaker	(1) rubber band	

Procedure: DNA Extraction

1. In a blender, mix a ratio of one cup of strawberries per one cup (250 mL) of distilled water. Blend for fifteen to twenty seconds, until the solution is a mixture. Blending or mashing the fruit with water causes some of the cells in the fruit to break open. Because DNA has a negative charge, it is able to dissolve in water. Many other cell parts will not be soluble in water.
2. In the 50 mL beaker, make a solution consisting of 5 mL of shampoo and two pinches of table salt. Add 20 mL of distilled water. Dissolve the salt and shampoo by stirring slowly with the stirring stick to avoid foaming. The cell membrane and nuclear membrane are broken down by soaps such as those found in shampoo and dish soap. When you wash dishes, the fats (grease) are removed from your dishes by the dish soap. When you wash your hair the shampoo removes the grease and oils. The positively charged sodium ions (Na +) are attracted to the negative charge of the DNA. This creates a "shield" around the DNA molecules and causes them to stick together (coalesce). This enables the DNA to precipitate out of the solution when added to alcohol in a later step.
3. To the solution you made in step 2, add 15 mL of the strawberry mixture from step 1. Mix the solution with the stirring stick for five to ten minutes. Avoid foaming.
4. While one member of your team mixes the strawberry solution, another will place the cheesecloth over the rim of the cup. Use a rubber band to secure the cheesecloth to the rim of the cup.
5. Filter the mixture by pouring it into the cheesecloth and letting the solution drain for several minutes until there is approximately 5 mL of the filtrate to test (it fills the bottom of the cup).
6. Filtering the soapy fruit solution through a coffee filter removes extra cell debris (cell membranes, precipitated proteins, and excess fruit pieces that didn't get completely pureed in the blender). The

> DNA molecules are soluble in the water, which enables them to pass through the filter.
> 7. Obtain a test tube of cold alcohol. For best results, the alcohol should be as cold as possible.
> 8. Fill the plastic pipette with strawberry solution and add it to the alcohol. DNA molecules are soluble in water BUT not in an alcohol solution. When the fruit DNA solution comes in contact with alcohol, the long, stringy DNA molecules precipitate into the alcohol. You can see this long, stringy precipitated DNA. What you see is thousands of DNA molecules that are stuck together.
>
>> FYI: Pure DNA is a colorless molecule. Any color that you may see in your DNA is caused by fruit pigments that got trapped in the stringy DNA.
>
> 9. Let the solution sit for two to three minutes without disturbing it. It is important not to shake the test tube. You can watch the white DNA precipitate out into the alcohol layer. DNA has the appearance of white, stringy mucus.

The plot of *The Gardener* revolves around the ethics of human genetic experimentation. Popularized by TV and movies, such as *X-Men*, *Jurassic Park*, *Spiderman*, and *Deadpool*, gene manipulation is a topic familiar to most learners; in fact, many students can easily explain that genetic manipulation requires "changing the DNA." However, when pressed further, most students have little understanding of what DNA looks like, at a macroscopic and molecular level. Extracting DNA from cells offers learners an opportunity to actually *see* genetic material. They can hold it in their hands, observe the translucent strands, and even take the DNA home with them. The experience takes an abstract concept from the pages of a textbook and brings it into reality and right into the learner's hands.

From this, students start to wonder, *if I can extract DNA from cells by following a simple protocol, how far-fetched is changing that DNA?* While students know intrinsically that changing the traits of an organism is complicated, a simple DNA extraction provides a launchpad for wonder. And, suddenly, the experience of the characters in *The Gardener* becomes relatable. The idea of TroDyn adding "an artificial element to boost the anomaly [that their subjects could maintain themselves without food or water] . . . artificial gene transfer," (p. 134) as Dr. Emerson explains to Mason in the novel, is more plausible.

Is It Ethical?

In the novel, the girl Mason meets, Laila, is one of several youths whose parents agreed to have their children as part of an experiment to "root them" and help them become able to live without food or water. In the process, they give the subjects a root system, made up of half organic material and half technological device, an advanced chip, designed to mutate the genes on a cellular level while connecting each of the children to each other, forming an artificial symbiosis that works like the real thing. (p. 137)

As Mason struggles to understand this, his "stomach turned," and he states, "They're playing God" (p. 135). Students, understanding that DNA is in all living things, can start to ask, *is that ethical?*

As Mason learns from Dr. Emerson after seeking her out for her history of working for TroDyn, the supposed reason for altering individuals such as Laila is to combat overpopulation and food shortage. He revels, "We couldn't adapt to something like starvation. We'd die from it first," and she responds, "it will take more than a few generations to develop the true autotroph. Which is why they wanted to employ an artificial element" (p. 137). Students might here consider the ethics of this, which teachers prompt: *If the population is really at stake, is it ok to engage in genetic modification? Where are the limits? To what extent should consent be part of this process?* The children involved, such as Laila, were not able to provide their own permission to participate. Instead, scientists sacrificed their offspring for this alleged greater good (p. 134). Students can consider if parents (and/or scientists) should be allowed to make decisions for their children if it will benefit science. Mason asks this same question, "experimenting on kids is so moral?" (p. 139) to which Dr. Emerson retorts, "For the greater good, I say it was. Their parents had the right to decide for them," (p. 139) and "it was the needs of the few against the needs of the many. The needs of the many will always be more important" (pp. 139–140). Students can debate their position and even dig further into parents' rights and scientific research.

Later in the novel, readers are told of more sinister desires for a self-sustaining army from the autotrophs. Eve, Laila's mother who has continued to work at TroDyn, is found to be "lobbying" to "make a deal with the military" (p. 213) and keeping parents in the dark about the plans for their children. Students can again consider the ethical implications here: *While a superhuman army that can exist without supplies could be a better defense for a country, are the sacrifices to humanity to create it justified?* Eve explains her motivation, stating that with the "power and financial backing" of the military, "this project could go leaps and bounds beyond" (p. 218). Still, however, readers should be prompted to consider if it should.

DNA Structure and Coding

It is human nature to look at something, be it robot, omelet, or the Eiffel Tower, and wonder, "How did they do that?" During reading, as students get to know the characters and their backstories, they will inevitably begin to wonder, *how* can the traits of a living organism be changed? This question is foreshadowed throughout *The Gardener*, especially as they learn how Mason's father, the Gardener, was modified. On page 195, Eve shares how they accomplished his transition, "We started substituting part of his blood with photosynthetic agents, all organic, derived from plants. As his body adjusted to a certain percentage, we increased it, until the photosynthetic agents outweighed the blood" (p. 195). This is a great line in the novel, it's impactful, horrifying, and intriguing all at the same time. Unfortunately, it's also bogus science. However, it presents teachers with the opportunity to do two things: (1) address the differences between science and science fiction and (2) capitalize on the moment to demonstrate to students how changes actually occur in living things—through genes and the code contained within DNA.

The molecular basis of DNA follows strict rules and by applying these rules, students can build models of DNA using any number of materials: paper, candy, construction bricks, and so on. A quick internet search will reveal numerous tried-and-true variations of this activity (e.g., Muskopf, 2021), so teachers don't need to start from scratch. The goal here isn't to inundate students with vocabulary and molecular chemistry; rather, the goal is to demonstrate the *patterns* that exist within a strand of DNA that is used to code for genes, that sugars and phosphates make up the backbone and the actual genetic code is expressed by the nitrogen bases.

From the DNA extraction activity, students should understand that living cells contain DNA. This activity builds on their growing understanding of the molecular nature of inheritance by demonstrating that the structure of DNA carries a unique genetic code. A code that, once one understands the nature of its structure and function, can be altered.

Teachers can guide students in making DNA paper models whereby students would "build" strands of DNA with color-coded paper pieces. It can be helpful to pair this activity with a video from Crash Course Biology to help students "see" the structure they are creating (Crash Course).

DNA PAPER MODEL

Introduction:

Imagine DNA as a twisted ladder. The outside of the ladder is made up of alternating sugar and phosphate molecules. The sugar is called deoxyribose. The rungs of the ladder are made of a pair of molecules called nitrogen bases. A sugar-phosphate-nitrogen base is called a nucleotide. There are four different bases in DNA: adenine, guanine, cytosine, and thymine.

Because of the chemical structures of the bases, adenine only pairs with thymine and cytosine only pairs with guanine to form a rung of the ladder.

Procedure:

1. From the paper provided by your teacher, color and cut out the chemical bases, sugars, and phosphates. Follow the directions in the table below:

    Table 8.1.

Molecule	Color
Sugar (Deoxyribose)	Red
Phosphate	Orange
Adenine	Yellow
Thymine	Green
Guanine	Blue
Cytosine	Violet

2. Arrange the cut-outs on your table to form the pattern described in the introduction. BE SURE YOU LAY ALL PIECES OUT BEFORE GLUING THEM TOGETHER! As a guide, you can attach the chemical base to the sugar molecule by matching up the dots. You can attach the phosphate group to your model by matching up the stars, and you can attach the top of the phosphates to the sugars by matching up the squares.
3. Glue the model together in your composition book.
4. Draw a color key next to your DNA molecule.

Conclusion Questions: (answer on a separate sheet)

1. What purpose does DNA serve in living organisms?
2. What are the three components of a nucleotide?
3. What is the name of the sugar molecule in the DNA helix?
4. What base does adenine pair with?
5. What base does guanine pair with?
6. Describe the shape of the DNA molecule. Draw a picture.
7. If the sequence of nitrogen bases on one DNA strand (one side of the DNA ladder) is AGCTCAG, what is the sequence of the bases on the opposite strand?
8. Assume that a 100-bair pair DNA double helix contains 45 cytosines, how many adenines are there?
9. An event occurs, resulting in the change of a guanine to a thymine. How does this *mutation* impact the genetic code?

A natural extension of this activity is to make changes, or *mutations*, to the models students build, to showcase how changes can be introduced (whether it be purposefully or by chance) into the genetic code. This also presents a prime opportunity to spend time exploring current events and news in science, including topics regarding CRISPR, using bacteria to produce compounds (such as insulin, biofuel, etc.), and genetically modified golden rice. If teachers want to return to the topic of science vs. science fiction, an extension is to explore counterexamples to *The Gardener*'s modification experience explained on page 195. Examples like chemotherapy, incorrect blood transfusions, or saline infusions show that change doesn't happen by injecting something into a person's body. Exploring these examples can help students understand the true nature of genetic change, through, genetic modification.

Baby Monsters

One final reading activity teachers might use is a coin-flip baby monster trait lab. Students will likely have completed something similar in earlier years, so the key to success with this activity is to build in levels of complexity (see figure 8.4 for a more complex version of the monster baby lab). Again, there are a lot of well-developed versions of this activity available via a simple internet search, but the focus should be on selecting a protocol (or building an extension) that presents students with the scenarios like those introduced in *The Gardener*.

To begin, students are provided a list of traits from parent organisms; some of the traits can be cosmetic, but there should be several that are essential for survival: color vision vs. visual acuity, feathers vs. fur, bipedal vs. quadrupedal, and so forth. There are numerous variations on how to complete the activity, so teachers can select a version appropriate for their specific learners. For a deeper dive, students should first identify parent genotypes, then follow the rules of inheritance (and a coin flip) to determine which alleles will be inherited by the offspring. This is a great option for more content exploration. For a less complex activity, a simple coin flip to select the trait being passed to the offspring will suffice. Once students have determined the traits of their baby monsters, they can draw and name their unique organism. This is where most activities end, but an extension will help students step into the shoes of the scientists in *The Gardener*.

Teachers can ask students how the creation of these baby monsters is similar to what the scientists at TroDyn had proposed to do to the children, for example. They can consider how long it would take for autotrophs like Laila to be fully developed and introduced into the world and why. Teachers can then propose a significant change to the baby monster's environment:

drought, famine, pollution, ozone depletion—whatever is selected, it should directly correlate to one of the inherited traits and affect survival. Students should recognize that switching a trait will help their baby monster survive. This is the perfect opportunity to propose the question, "Is it okay to change an organism's traits?" Sometimes it helps to create false gravity by linking the survival of the baby monster to a passing grade or extra homework assignment (the key here is *false* gravity, there's no real link to a grade or extra work). If students have buy-in to their baby monster's survival, does that make it okay to change their traits?

This line of questioning clearly connects to *The Gardener*. In the novel, as described above, Dr. Emerson shares with Mason the TroDyn scientists' task of the survival of future generations in a world plagued by famine and drought. Young adults may immediately identify with the struggles of the main character, who often feels powerless trying to navigate a world full of rules and adults making decisions. The goal of the baby monster trait activity is to guide their thinking past Mason and Laila's immediate experiences and to consider the perspective and reasoning of the scientists, which is to "do something to alter the path" (p. 189) of a dying planet rather than giving in. Ultimately, playing with the concepts of genetic code and manipulation while reading *The Gardener* invites students to meaningfully explore the questions presented over the course of the story and from multiple perspectives until they are ready to address the ultimate ethical ponderance, *"is human genetic experimentation okay?"*

MONSTER BABY LAB: MORE COMPLEX VERSION

Instructions: Pretend that monsters are real, and they take on the same basic genetic rules as humans.

1. With a partner, decide who will be monster parent 1 and who will be monster parent 2. You're going to make a baby monster!
2. Each monster parent will flip a coin for each trait. The trait will be made up of two genes and designated by one letter. The exception is when a trait has many variations, or alleles, such as eye color. In this case, each parent will be asked to flip the coin twice for two different letters.
3. **Heads** is dominant. **Tails** is recessive.
4. Fill out the chart with your partner, starting with the face shape of your monster and ending with the gender. Once you have determined

the genotype and phenotype of your baby monster, you will draw it on paper. It should fill up the whole paper!
5. Your chart will have five columns (trait, monster parent 1 gene(s), monster parent 2 gene(s), genotype, and phenotype) with the following listed under the traits column: body shape (D), number of eyes (E), shape of eyes (S), eye color (A & B), mouth shape (M), monster fur color (F & G), monster fur pattern (H), pattern of fur color (F and G), and gender. Here is a sample below demonstrating how your chart should look when you fill it out. Results will vary depending on the coin toss.

Table 8.2.

Trait	Monster Parent 1 Gene(s)	Monster Parent 2 Gene(s)	Genotype	Phenotype
Monster Body Shape	D	D	Dd	Round
Monster Eye Color	AB	AB	AABB	Red

6. Gender will be the last row on the chart. Monster mom always contributes an X and that is already filled in. Now dad will determine the gender by flipping a coin. If it lands on heads it is a Y and tails is an X.
7. Once the coin toss is complete and the chart is filled in, the next steps will be to DRAW, COLOR, and NAME your monster baby. Add some arms and legs to finish it up. Make your drawing amazing!

Monster Traits

- **Body Shape:** Round (D) is dominant to Asymmetrical (d).
- **Number of Eyes:** Two Eyes (E) are dominant to one Eye (e). Incomplete dominance results in three Eyes.
- **Shape of Eyes:** Round (S) is dominant to Almond (s).
- **Color of Eye(s):** Use two letters and each parent flips the coin once for each letter.

Table 8.3.

AABB—red	AaBB—yellow	aaBB—blue
AABb—pink	AaBb—brown	aaBb—purple
Aabb—orange	Aabb—green	Aabb—black

- **Mouth Shape:** Smile (M) is dominant to Snarl (m). Incomplete dominance results in a frown.

- **Monster Fur Color**: Use two letters and each parent flips the coin once for each letter.

    Table 8.4.
FFGG—blue	FfGG—orange	ffGG—yellow
FFGg—pink	FfGg—green	ffGg—gray
FFgg—purple	Ffgg—brown	Ffgg—gold

- **Pattern on Fur**: Stripes (H) are dominant to No Pattern (h). Polka dots result from codominance.
- **Pattern on Fur Color**: If your monster baby has a fur pattern from number 7, go back to number 6 to determine the color of the polka dots or stripes. Using letters F and G once again, record the genotype and phenotype on your chart (there's an extra space at the bottom for this). If your monster has no pattern, skip this step and move on to number 9.
- **Gender**: The monster dad will now determine the gender of the baby by flipping only one coin. Remember, the monster mom has already contributed an X. If the coin lands on heads it is Y and tails it is X. XX—girl. XY—boy.
- With a complete chart, it is time to DRAW, COLOR, add arms and legs, and NAME your baby monster.

Helpful Vocabulary:

- **Allele**—alternative forms of a gene for each variation of a trait of an organism.
- **Codominance**—a form of dominance where the alleles of a gene pair heterozygously and are fully expressed. This results in offspring with a phenotype that is neither dominant nor recessive.
- **Dominant**—observed trait of an organism that masks the recessive form of a trait; expressed with a capital letter, that is, "B."
- **Genetics**—branch of biology that studies heredity.
- **Genotype**—combination of genes in an organism.
- **Heterozygous**—when there are two different alleles for a trait (i.e., Bb).
- **Homozygous**—when there are two identical alleles for a trait (i.e., bb or BB).
- **Incomplete dominance**—a form of inheritance where one allele for a specific trait is not completely expressed over its paired allele. This results in a third phenotype where the expressed trait is a combination of the phenotypes of both alleles.

- **Law of independent assortment**—Mendelian principle states that genes for different traits are inherited independently of each other.
- **Law of segregation**—Mendelian principle explains that because each plant has two different alleles, it can produce two different types of gametes. During fertilization, male and female gametes randomly pair to produce four combinations of alleles.
- **Phenotype**—outward appearance of an organism, regardless of its genes.
- **Recessive**—trait of an organism that can be masked by the dominant form of a trait. Expressed as a lower-case letter, that is, "b."
- **Trait**—characteristic that is inherited; can be either dominant or recessive.

Question Ball Toss

The Gardener clearly lends itself to discussing ethics, morals, and choices. Teachers may want to check in with students regarding their thoughts throughout reading the text. With potentially difficult conversations on complex topics, it is important for teachers to provide small structured opportunities for students to share opinions in a safe space. An effective, simple activity for sharing opinions aloud is a Question Ball Toss. Soccer balls, beach balls, and large bouncy balls work great for this activity. Regular check-ins are important for building a classroom culture supporting conversation and dialogue. This activity can offer a change of pace after reading several chapters or could work well as an end-of-week reflection. Teachers can either write questions directly on the ball or use numbers that correspond to a revolving cache of questions posted on the whiteboard, screen, or around the room (or student-generated questions). Students toss and catch the ball—whatever number or question their thumb is closest to; they answer aloud before passing the ball along to the next student. The complexity of student answers will vary depending on their comfort level speaking aloud and/or sharing, so it is important to be mindful of individual student needs, participate in the activity, and model the active listening skills students should aim to grow. Activities like this can help students see not only the connection between the text and ethical concepts but also can give space for all opinions to be shared.

This activity can be completed at various stages of the novel, as Mason learns more and wrestles with his own conscience. For instance, teachers can use the questions in table 8.5 to coincide with ethics-raising instances in the book.

With each, students might again need to engage in dialogue to share and weigh opinions and further develop their own stance on this and other ethical issues.

Studying Genetics and Ethics through Young Adult Literature

Table 8.5. Examples of Discussion Questions Related to Textual Topics

Topic/Issues	Textual References	Suggested Questions for Discussion
Mason learns what is being done to Laila	"Mom's head moved from side to side, almost slow motion as she continued to stare at the girl. 'She's not supposed to wake up.' Her voice lowered to a whisper. 'They aren't ever supposed to wake up.'" (p. 41) "I've never seen such gnarled, nasty scars." (p. 58)	How are Laila's rights being violated?
Mason uncovers from Dr. Emerson the reasons behind creating autotrophs	"You must understand I truly was passionate about finding an answer to starvation. I mean, to actually be able to count yourself as one of the scientists who solved a crisis like that? Especially when the future will see all of us struggling for food during climate change." (p. 131)	To what extent were Dr. Emerson's actions in helping at TroDyn justified? What are the limits to these actions?
Eve's plan to create a military is revealed and executed	"'The deal is done. Once I took care of Solomon . . .' She glared. 'Let me rephrase. Once I thought I took care of Solomon, I called my contact at the military. They'll be here in twenty four hours.'" (p. 223)	What would be the advantages of a self-sustaining army? Should certain individuals be sacrificed for the greater good?

AFTER READING *THE GARDENER*

Revisiting the Formative Probe

Upon finishing the novel, it will be important to revisit the formative assessment probe that launched the unit. Provide students an opportunity to look at the scenario again, this time with a deeper understanding of the molecular nature of inheritance and genetics. Repeating the same activity is always an option, but it can be useful to add new elements, so students aren't bored. In addition to asking students to pick a response with which they most agree and justifying their choice, teachers can, for example, have students review their initial choice, this time agreeing or disagreeing with cited evidence. It is useful to include opportunities for metacognition; questions such as, *"What do you think now? Has your opinion changed? Why? How?"* can be valuable tools to help students practice expressing and justifying their responses. A simple extension is to pose additional questions such as, *"What if this organism was in danger of extinction? How would we create a solution for that? Should we create a solution for that?*" Questions like these directly relate to *The Gardener* and the threat of famine, "a universal problem," (p. 139) and provide students a safe space to begin crafting their own educated opinions about how far science should go to address social issues.

Let's Talk

Teachers might also guide students in planning and participating in Socratic seminars regarding topics connected to both the novel and the scientific concepts introduced in the unit. There are numerous protocols for Socratic seminars; the key is to select one that allows students to freely express their justified opinions, actively listen to the reasoning of others and practice/participate in constructive dialogue. For example, Gonzalez (2015) offered a "Big list of class discussion strategies" on the Cult of Pedagogy website that could be very useful for teachers who want new ideas on ways to set up and facilitate class discussions.

When introducing and facilitating class discussions, it can be helpful to team students into smaller dialogue communities and explicitly build conversation norms—all practices that are seen through global scientific communities. Teachers can then gradually release responsibility to students and abstain from answering the questions themselves; student growth occurs through repeated conversation and consideration with their peers.

The issues presented in *The Gardener* are real ethical quandaries that are addressed across the field of science daily. Students can revisit the ethical questions posed in the "Introducing Ethics" section of this chapter, explaining if, why, and/or how any of their opinions have changed. Teachers can then lead students to consider other current ethical issues outside of those presented in *The Gardener*. For example, teachers might have students consider how the environment changes temperature, using examples of polar bears and melting ice caps or tree moths during the Industrial Revolution. They might also discuss how when a prey species goes instinct, food sources change, such as in the case of the Harelip suckerfish and snails. Finally, they might consider how when an invasive species arrives, they might serve as a new predator or a direct competitor for food, such as with the Sea Lamprey in the Great Lakes.

Ultimately, there is no "right" or "wrong" thinking in this regard; however, it is essential that students build a practice of using evidence as a basis for their opinions and conclusions. Students can benefit from expressing their own reasoning and hearing the justification for differing opinions. Similar to the goals of *math talks* or *literary circles*, Socratic seminars aim to provide students with a safe place to explore different ideas, approaches to reasoning, and opportunities for expression. If teachers want to just stay focused on *The Gardener*, some critical questions to ease students into dialogue include:

- How was Mason helping Laila? How was he hurting her?
- How can the scientists' actions be understood, if at all?
- How might genetic engineering be a viable solution to food shortages around the world?
- Why might someone choose to change the traits of an organism?
- In what ways is human genetic experimentation ethical? In what ways is it not ethical?

In all cases, it is important to ease students into the conversation, starting with questions (similar to the formative assessment scenario) that are low risk and ending on the penultimate ethical dilemma.

EXTENSION ACTIVITIES

Working with *Drosophila*

There are a multitude of ways to extend student thinking and learning regarding genetics and ethical science practices. *The Gardener* presents an opportune stage to explore the molecular nature of inheritance further or to make a leap and start to explore the principles of evolution. If teachers are interested in diving deeper into the molecular nature of inheritance, a powerful extension activity is to work with *Drosophila* (Lakhotia, 2021). Also known as the fruit fly, *Drosophila* is a great model organism for studying traditional Mendelian genetics. There are numerous laboratory protocols available for free, like the Biology Corner (Muskopf, 2022), NCSS Biology (2010–2022), and HHMI: BioInteractive (n.d.). *Drosophila* cultures can even be purchased from large online retailers. If teachers wish to avoid working with living organisms, there are online simulations that can easily be accessed from the above or other websites. Studying *Drosophila* can be a launchpad for any number of extension studies, ranging from traditional Mendelian genetics to sex-linked traits to mutations.

Bird Beak Lab

Another possible extension activity that leads to further discussion of adaptation and evolution, which are inherently related to the principles of genetics, is the Bird Beak lab (BioInteractive, n.d.). In general, students use different utensils (forks, spooks, chopsticks, clothespins, etc.) to collect food from a dish. The different utensils are models for different types of bird beaks and the dish (filled with beans, rice, rubber bands, etc.) represents food in the environment. Teams of students collect food over several one-minute rounds, documenting how much of each type of food is collected. In the end, the teams contribute their findings to class data to be interpreted. It becomes fairly obvious that certain types of "beaks" are best suited to specific foods, so the follow-up is to change the type of food available in the environment and repeat the exercise. This activity quickly illustrates the inciting event of *The Gardener*: organisms that cannot adapt to environmental change will not survive (p. 136). This activity also can serve as a segue into broader conversations regarding inheritance, traits, and the fundamentals of evolution. It may be appropriate to reintroduce the main question from *The Gardener* in this slightly different context: "*Is animal genetic experimental ethical?*"

Lab Experiment

Students could, for example, develop their own hypothetical lab experiment regarding growing a hard-to-kill-fast-growing plant, making connections to the autotrophs developed in *The Gardener* (p. 195). Then students could argue for or against the government mandating that people grow and/or eat this plant. An activity like this, again, returns to the question of: *If* we can do something, *should* we? What are the pros and cons? Who will benefit? Is it ok to make governmental mandates if it is to help vulnerable populations?

Social Action Projects

The ethics of genetic research is a largely misunderstood area in public discourse. To combat this, and to have students explore and engage in their own communities, teachers might assign the development and implementation of social action projects (Boyd & Darragh, 2019). In such work, students target a specific problem related to an issue and develop a related action. Teachers could require action projects focused around a topic of their choice that is found in *The Gardener*. For instance, they could develop awareness campaigns around what is actually happening in genetic research today and both its benefits and possible detriments. They could hold an informative event for their school or broader community as well, disseminating their research and even providing recommendations for future endeavors in this realm.

Ethics Research Topics

Another related and quite a relevant area students could research is embryonic engineering. Students have likely heard of IVF and other methods for addressing fertility issues, but an opportunity to unpack and weigh the ethics of such efforts could really expand their learning from the text. They could again create informative materials to spread awareness. They might research couples who have spoken out about taking this route and consider the ethics of such individuals' decisions, specifically reading their personal narratives. Because such scientific efforts are often governed by legislation (e.g., the ban on stem cell research in the 2000s), students might also invite legislators to speak on how they make decisions around such entities. Especially when religion and faith intersect with scientific decisions of this nature, it will be helpful for students to understand the various facets that govern such decision-making.

Community Book Club

Students might also lead a community book club in which they invite parents, administrators, and other community stakeholders to read *The Gardener* and discuss its implications. Youth could serve as facilitators of the book clubs and devise a written guide in addition to questions to solicit opinions and thoughtful dialogue on the book and its topics. As Mason's mother had specific reasons for her choices in the text, parents might weigh in on related topics. She had fallen in love with the Gardener, but when she became pregnant, she left. She didn't want anyone to know about her child or worse, as Mason's father's shared, to "force her to hand you over" (p. 178). Doing what is best for children, especially separating them from a parent or loved one, even if in their best interests, can be difficult. Parents could discuss this and apply to current contexts.

Creating a Podcast

To address issues in the book related to food and climate, students could design and host a podcast series in which they interview local farmers and climate change experts, uncovering what concerns are most pressing in their areas and what factors are most influencing their supplies. They might then problem-solve ways to address any imminent issues with food supply and find ways to advertise those in their schools and even implement those on a small scale, such as planting trees and reducing their carbon footprints.

There are a number of resources available for helping students create podcasts. National Public Radio, for example, has a "Starting Your Podcast: A Guide for Students" (2018) that walks a person through getting started, planning, and producing a podcast. They cover the materials needed as well as suggestions for free software that can be used. Often, students' phones are the easiest and most accessible means for capturing audio. This site also includes examples of podcasts, tips for structuring them, and even details on designing a solid interview and script. Lamb and Johnson (2007) also created a brief overview for teachers creating podcasts that they could use to direct students; they instruct creators to "look for activities where audio adds a dimension that would not be available with another medium, such as the intensity of voice found in a commentary or an interview or in storytelling or oral music" (p. 61) and emphasize the importance of audience when creating a podcast.

CONCLUSION

There are clearly a multitude of ways in which teachers can use *The Gardener* to encourage student engagement in not only biology but also in their schools, their communities, and their world. The concepts of DNA and genetic engineering are often difficult for students to understand, and yet they are all around them in their own lives as well as in popular media. Mason's experience with Laila in the text provides a storyline through which to first understand these theoretical ideas and then to consider the difficult ethical decisions that can accompany them.

Through a fictional case, *The Gardener* shows how individuals can be genetically modified supposedly for the greater good. Students can determine where the boundaries of science should lie when the health and welfare of humans are at stake—when Laila becomes not a person, but a robotic entity living in a controlled environment (p. 205). By studying how *The Gardener* created autotrophs (p. 195), students can learn more about not only genetics but also how sinister motives affect creation. As such, this study can help students see science as more than a static discipline with *right* or *wrong* answers but instead as a field in flux that changes with modern developments and the discovery of new knowledge.

REFERENCES

BioInteractive. (n.d.). *The origin of species: The beak of the finch.* https://www.biointeractive.org/classroom-resources/origin-species-beak-finch.
Boden, S. A. (2010). *The gardener.* Feiwel and Friends.

Boyd, A. S., & Darragh, J. J. (2019). *Reading for action: Engaging youth in social justice through young adult literature.* Rowman & Littlefield.

Crash Course. (n.d.). *DNA structure and replication: Crash course biology #10.* https://www.youtube.com/watch?v=8kK2zwjRV0M.

Gonzalez, J. (2015). The big list of class discussion strategies. *Cult of Pedagogy.* https://www.cultofpedagogy.com/speaking-listening-techniques/.

HHMI: BioInteractive. (n.d.). https://www.biointeractive.org/classroom-resources.

Keeley, P. (2021). *Uncovering student ideas in science: 25 more formative assessment probes.* (2nd ed.). NSTA Press.

Lakhotia, S. C. (2021). *Experiments with Drosophila for biology courses.* https://bdsc.indiana.edu/pdf/Experiments_with_Drosophila_for_Biology_Courses.pdf.

Muskopf, S. (2021). *DNA-build it.* The Biology Corner. https://www.biologycorner.com/2021/02/15/dna-build-it/.

NGSS Biology. (2010–2022). https://www.ngsslifescience.com/science.php/science/biology_lesson_plans.

NSTA. (2007). *Baby mice 129–136.* https://www.biologycorner.com/2021/02/15/dna-build-it/.

Science Buddies. (2013). Squishy science: Extract DNA from smashed strawberries. *Scientific American.* https://www.scientificamerican.com/article/squishy-science-extract-dna-from-smashed-strawberries/.

Chapter 9

Hungry for More

Exploring, Experimenting, and Engineering with The Hunger Games

Leslie Suters and Kristen Pennycuff Trent

The first novel in *The Hunger Games* trilogy by Collins (2008) provides a rich context for exploration in Life Science classrooms. The novel includes multiple entry points for actively engaging students in three-dimensional instruction focused on disciplinary core ideas, science and engineering practices (SEPs), and crosscutting concepts (CCCs) as recommended by the *K-12 Framework for Science Education* (National Research Council, 2012). This chapter offers approaches for ways to develop scientific literacy through the lens of critical reading and technical writing focused on themes aligned with meaningful current events and topics of interest to adolescents.

Life Science topics that are prevalent in *The Hunger Games* novel include genetically modified or engineered organisms such as the creation of birds that can repeat human conversations used as spies and wasps and wolves used as weapons. A genetically modified or engineered organism (GMO) has been changed by adding to or removing DNA sequences to produce a desired trait in a way that would not occur naturally. Common real-world uses of genetic engineering include modifications to plants to make them easier to grow and/or to include additional nutrients. Animals, including cattle, pigs, chickens, rats, and mice, can also be genetically modified, and in some cases, these animals are used for human consumption. Genetic engineering research has focused on curing diseases such as cancer and producing items such as human insulin, vaccines, and disease-resistant plants. While there are many positive aspects of using genetic engineering, there are concerns about the ethics and risks associated with changing the natural makeup of organisms. There are apprehensions about the negative side effects to the GMO itself and to those that use or are ultimately impacted by the GMOs. By creating GMOs with specific traits, we could be limiting genetic diversity, reducing nutritional values, producing unknown side effects, and interfering with the naturally occurring ecosystem. The use of genetic engineering leads to important discussions of bioethical issues and legislative controls for when it should be used and how it should be regulated. Technological innovations such as Clustered Regularly Interspaced Short Palindromic Repeats (CRISPR) have made it easier and less costly to edit genes through removal, additions, or silencing of specific DNA sequences since its inception in 2012.

The Hunger Games also includes a number of rich connections within the realm of ecology. Each character has a unique position within the world of Panem and within the district in which they live. They have different skills and attributes that help them survive whether through hunting and determining what is safe to consume, using camouflage as a means of survival, collecting plants that can be used for healing, or knowing that it's important to locate potable water and how to safely purify water. As such, reading this book in the science classroom affords students an opportunity to explore ecosystems (predator/prey, camouflage, and teamwork), survival skills focused on living off the land (edible mushrooms and plants, hunting, and balanced diets), medicinal herbs (apothecaries), and water filtration.

THE HUNGER GAMES BY SUZANNE COLLINS

The Hunger Games is a dystopian young adult novel set in Panem, formerly the United States, ruled by President Snow who resides in the Capitol. The majority of the citizens of Panem live in one of twelve districts which have different jobs to support the citizens who live in the Capitol. As one of many ways to control the districts, every year the Capitol has required each district to send one girl and one boy between the ages of 12 and 18 as tributes to fight to the death in "games" until only one tribute is left. This book catalogs the seventy-fourth Hunger Games. The major characters in the book are the two tributes from coal-mining District 12 including sixteen-year-old Katniss Everdeen, who has strong survival skills from living off the land, and Peeta Mellark, who has artistic and interpersonal skills. Haymitch Abernathy, a former District 12 Hunger Games champion, is their mentor.

PREPARING STUDENTS TO READ
THE HUNGER GAMES

Developing Student Opinions about Bioethics, Genetics, and Ecology

Anticipation Guides have been described as "a strategy in which students forecast the major ideas of a reading passage through the use of statements that activate their thoughts and opinions" (Yell & Scheurman, 2004, p. 361). Prior to reading the novel, students can individually read statements related to genetics, ecology, and associated bioethics concepts central to the text and indicate if they agree or disagree and why. Next, they meet in small teams to discuss their answers and rationale, as well as to negotiate a consensus for each statement. During reading, students can use the Anticipation Guide to take notes or list evidence that corroborates or refutes their opinions. After reading, students can return to the guide, completing a Reaction column for the statements that applied to that segment of the text, along with textual evidence that justified their response. Students once again meet with their teams to share any changes, explain their thinking, and again reach a consensus on each statement. Figure 9.1 presents statements for an Anticipation/Reaction Guide representing both bioethical issues and formative probes to uncover misconceptions in science concepts.

Using Vocabulary to Make Predictions

The 6S Strategy is a pre-reading strategy based on Splash-Sort-Label (Rollins, 2014) that requires students to Splash, Sort, Scribe, and Support followed by Stay and Stray for reporting results. Students are first presented with vocabulary words (such as those suggested in figure 9.2) both from the novel and related science concepts. To begin, separate students into districts—replicating the setting dynamics in the book. Working in a team with

Directions: Read each statement below then circle YES if you agree with the statement and NO if you disagree. With tablemates, discuss responses and reach a consensus, circling YES/NO. Respond again, after reading the book, using text evidence to support your responses.

SELF		CONSENSUS	After Reading	Text Evidence for Response
Yes/No	Hunting is cruel.	Yes/No	Yes/No	
Yes/No	Having children is a blessing.	Yes/No	Yes/No	
Yes/No	People have the right to complain about their government.	Yes/No	Yes/No	
Yes/No	All plants are edible.	Yes/No	Yes/No	
Yes/No	Camouflage can only occur naturally, such as a moth blending into a tree.	Yes/No	Yes/No	
Yes/No	Alterations to genetic code can only occur through evolution.	Yes/No	Yes/No	
Yes/No	It's not safe to use herbs you find in a forest as medicine.	Yes/No	Yes/No	
Yes/No	Scientific innovations are helpful to humans and the environment.	Yes/No	Yes/No	

Figure 9.1. *The Hunger Games:* Anticipation/Reaction Guide Prompts.

BOOK PART	SCIENCE CONCEPTS	
	Term	Pg. #
Part I The Tributes	predator pelt apothecary prey genetically altered muttations & mutts synthetic (fire) camouflage variables	4, 11 6 8 11 42 42 67 95 185
Part II The Games	catacombs adrenaline camouflage evolve foliage canopy edible herbal stamen blossom nectar reptiles	143 151 153, 154 152 164, 183 165 180 188 196 196 196 288
Part III The Victor	variable venom (rocky) terrain Blood poisoning yearling carcass hypodermic (peaceful) terrain	250 250 251 265 269 269 289 325

Figure 9.2. *The Hunger Games* Science Vocabulary.

other members of their district, students can perform an open sort by splashing the images and words on the table in front of them and working together to arrange the words into meaningful categories with heading cards that they create. From the sort, they then either work with their district or as individual tributes to scribe a detailed paragraph predicting what the text will be about and incorporating as many of the images and words as possible. As they finish, they must first justify their categories.

Once all districts have completed their sorts, have them share with other teams using a Stay and Stray strategy. One tribute stays with the completed project to support and explain their thinking to visitors from other districts, while the other members stray to visit different districts to hear their rationale. Team members rotate roles so everyone has the opportunity to be the one who stays and is part of the team that strays. Figure 9.3 includes graphics from Photos for Class (2022) and illustrates a completed 6S Strategy for "ecosystems." In this example, teams were provided digital images with captions of science terms and a list of other terms from the text for sorting.

Hungry for More 161

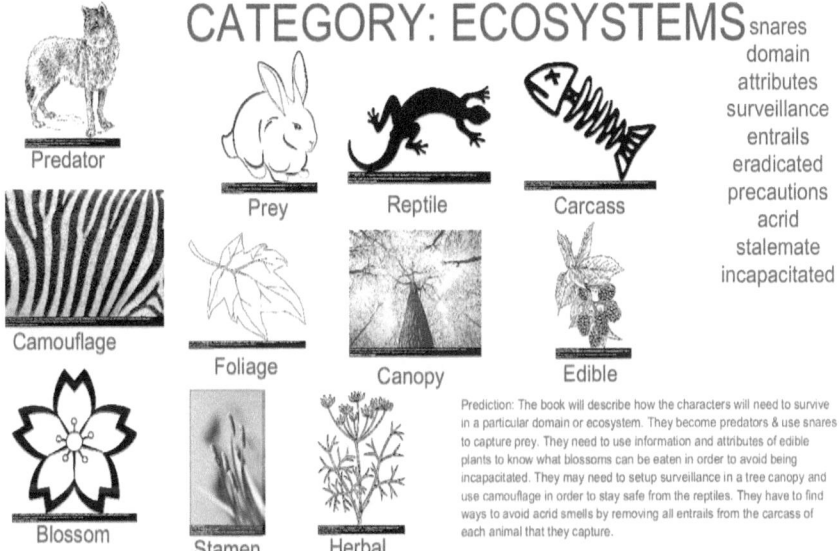

Figure 9.3. 6S Strategy (Splash, Sort, Scribe, Support, Stay, and Stray) Example.

WHILE READING *THE HUNGER GAMES*

Exploring Biotic and Abiotic Resources within Panem

While reading *The Hunger Games*, students use a Semantic Feature Analysis (Johnson & Pearson, 1978), or SFA, to explore the relationships between and among the biotic or natural and abiotic or geopolitical resources available in each of the major settings: District 12, the Capitol, the training center, and the arena, as well as the impact those resources has on the characters and the development of the plot. While this strategy typically has students signify the relationships with basic symbols such as a plus sign (+), a minus sign (−), or a question mark (?), more robust analysis would require students to cite textual evidence for the relationship or to use a Likert-type scale where 4 represents a strong amount of evidence for that resource, 2 indicates little evidence, and 0 signifies no evidence in the text (Johnson & Pearson, 1978). Completing the graphic organizer is important, however, the true power of this strategy lies in the discussion between students as they justify their ratings or evidence and reach a consensus. As students complete the organizer while reading and discussing the results collaboratively, they are activating background knowledge, determining hierarchical relationships, and confirming or building and integrating their knowledge of science and ELA concepts. Table 9.1 outlines a sample SFA that integrates the Likert-type response with textual evidence.

Table 9.1. Semantic Feature Analysis for Natural and Geopolitical Resources across Settings

Natural or Geopolitical Resources	District 12	Capitol	Training Center	Arena
Landscape	4	1	1	4
	scruffy field called The Meadow; surrounded by high chain link fence with barbed wire coils on top; forest across fence; many coal mines; Hob (black market) in abandoned warehouse; hills; rock ledge; public square bordered by a few shops; oak trees	broad avenues; skyscrapers that glisten and are in a rainbow of colors; oddly dressed people with bizarre hair and painted faces who have never missed a meal; Remake Center; City Circle; President Snow's Mansion	rooftop deck and garden surrounded by electrical shield	lake; pine trees; valleys; willow trees; bloodred berries; pond lilies; spring-fed pools; shallow hole filled with leaves; wildflowers; muddy creek banks; boulders; small cave-like structures; field with grasses shoulder high in patches of different colors; ponds; plains; night lock berries
Wildlife	4	0	0	4
	wild dogs; cougar; bear; lynx; snakes; deer; fish; rabbit; squirrel; mockingjays; turkey	N/A	N/A	tracker jackers; mutations; rabbit; fish; deer; wild dogs; owl; groosling; skunk
Edible Plants	4	0	0	2
	basil; dried mint; apples; berries; greens; strawberries; wild plums; Katniss plants	N/A	N/A	bloodred berries; edible water plants; starchy root-like parsnip that Rue contributes; nuts; greens
Electricity	2	4	4	0
	"comes and goes; usually a few hours every day; only reliable for airing Games or government message on TV that's mandatory"	"The Capitol twinkles like a vast field of fireflies" (p. 81). There would be no shortage here; ever	electric field on roof deck to prevent suicide	current binds tributes to the hovercraft; electricity for game makers to use but not tributes

Healthcare	2 apothecaries; natural remedies; medicinal herbs	4 miracle cures; prosthetics; plastic surgery and anti-aging treatments; teams of doctors; deafness and scars healed	4 staffed hospital underneath gym	1 bottle of iodine; sponsor gift of ointment; leaves that heal tracker jacker stings from Rue; fever-reducing medicine from first-aid kit; sponsor gift of sleep syrup; hypodermic needle (antibiotics)
Food	2 coarse bread from tessera grain; oil from tessera; flora and fauna from woods; bakery bread for special occasions; Greasy Sae's Winter Specialty: mice meat; pig entrails; and tree bark; dandelion greens; pokeweed; wild onions; pines; wild turkey; Katniss tubers boiled or baked like potatoes; honey	4 sugar; hot chocolate; rolls shaped like flowers; carrot soup; green salad; lamb chops; mashed potatoes; cheese; fruit; chocolate cake; coffee; eggs; ham; fried potatoes; orange juice; hard candies; pudding the color of honey; chicken; cream; green peas and onions; pearly white grain; wine; mushroom soup; bitter greens with tomatoes the size of peas; rare roast beef; noodles in green sauce; cheese; sweet blue grapes; cake; eggs; sausages; pancakes with orange preserves; purple melons; beef stew; roasted pig with apple in mouth; lamb stew; dried plums; wild rice; cream and rose petal soup	4 clear broth; applesauce; water; roast beef; peas; rolls;	1 a pack of crackers; dried beef strips; rabbit; groosling; waterbird eggs; roots; berries; greens; apples; crescent roll from District 11; sponsored hot broth; stew made of groosling and Rue's roots with chives; sponsored feast after second kiss of fresh rolls; goat cheese; apples; lamb stew and wild rice

Rating Scale: 4 = Excellent Evidence; 3 = Very Good Evidence; 2 = Fair Evidence; 1 = Poor Evidence; 0 = Very Poor or No Evidence Charting Character Actions and Responses to Events and Environment.

Students can create character maps charting the trajectory of each character as they read *The Hunger Games*. These can be kept as handwritten or electronic journals that they revisit after each chapter. Figure 9.4 illustrates a journey of the high and low points of events in Parts I and II of the book associated with the character Peeta Mellark. Individual or small teams of students could be assigned different characters to track and multiple teams can even map the same characters to provide discussion points as to what events are selected and why they were rated certain scores. This will help bring in differences and similarities in personal connections and/or background knowledge that individual students associate with the events in the text. For example, in Part II of the novel, Peeta allies with the Career tributes during the Games and Katniss overhears them talking about killing her (p. 161). Figure 9.4 shows this instance rated at a −2 due to how hard it was for Peeta to pretend; however, some students might rate this as a +2 because Peeta was actively trying to protect Katniss and that made him feel worthwhile during one of the toughest times of his life.

Guiding Discussion of Ecological Concepts

As a way to encourage students to think deeper about the science concepts presented within *The Hunger Games*, ask them to identify textual evidence

Character Journey For: Peeta Mellark Select important events chronologically from the book for your character and rate them using the following scale: -3=hopeless; -2=depressed; -1=anxious; 1=hopeful; 2=happy; 3=elated	
Chapter 2 p. 25 - The Reaping Peeta is selected as the male tribute from District 12.	-3
Chapter 5 p. 69 - Opening Ceremony Matching outfit with Katniss and they hold hands on chariot and wave to the crowds.	1
Chapter 6 p. 80 - Rooftop at the Training Center Peeta invites Katniss to the rooftop so they can have some privacy. Katniss shares memory of seeing Avox girl in the woods of District 12 running and then being captured by a Hovercraft.	1
Chapter 7 p. 95-96 - Training Center Excels at camouflage station - mixing mud, clay, & berry juices to weave disguises from vines & leaves on his skin.	1
Chapter 8 p. 108 - Training Score Peeta scores an 8 from showcasing his strength to the Gamemakers.	3
Chapter 9 p. 130 - Interview with Caesar Flickerman Peeta tells Caesar he has a "crush" on Katniss. Helps establish his winning personality and create interest for watching Peeta & Katniss during the Games.	2
Chapter 10 p. 140 - Rooftop at the Training Center Peeta tells Katniss that he doesn't want the Games to change who he is. "I want to die as myself." Katniss tells Peeta, "Care about what Haymitch said. About staying alive."	-1
Chapter 12 p. 161 - Peeta allies with the Careers Katniss overhears Peeta talking with the careers who are hunting for her.	-2
Chapter 14 p. 193 - Peeta tells Katniss to run After the tracker jacker attack, Katniss, Peeta and the Careers are injured. Katniss and Peeta run into each other and Peeta tells her to run. Soon after Cato injures Peeta with his sword.	-3

Figure 9.4. Character Journey for Peeta Mellark—Parts I and II of *The Hunger Game*.

they can use to compare and contrast Katniss' and Peeta's life experiences in District 12 and how these experiences impacted survival during the games. As students gather and assimilate this evidence they can use the information to sketch and annotate models of the ecological systems of both District 12 and *The Hunger Games* arena. As students read, the teacher could pose the prompts/questions below to guide them. Table 9.2 provides page numbers of textual evidence.

- In the context of District 12, describe the resources Katniss and Peeta used to survive.
- Note and describe the resources the tributes use to their advantage to survive.
- What were the limiters for tributes?
- How might the tributes' performance change if the arena was in a different ecological system?
- How do characters treat injuries? What medicine do they have within District 12? The arena?
- Describe the predator/prey relationships within District 12. Explain how these relationships compare and contrast to those in the arena.
- What ideas in the text support and validate teamwork as a means of survival within their districts? Within the arena?

AFTER READING *THE HUNGER GAMES*

Connecting Content and Crosscutting Concepts

Hexagonal Thinking encourages students to think critically, make claims that are supported by evidence, and determine relationships between and among important ideas with a concrete model. While it can be used as a before and during reading strategy, it may be most effective after students have read the entire novel so that they can make use of many connections they have made within the text as they are reflecting.

After reading, students can be presented with multiple hexagons labeled with major vocabulary terms, science concepts, and story elements such as characters and themes from the novel. Blank hexagons should also be included for students to add other thoughts and ideas as they work in small teams to sort the hexagons into a meaningful arrangement. Each hexagon can be connected to up to six other hexagons, and students must be able to justify why each piece is placed in that particular location. They should also be encouraged to try positioning hexagons in a variety of different places to achieve the most meaningful representation possible. It is also important

Table 9.2. Katniss and Peeta's Life Experiences and Survival

	Katniss		Both		Peeta	
Traits for Survival	District 12 and Training Center	Arena	District 12 and Training Center	Arena	District 12 and Training Center	Arena
Learned Lessons from Childhood	pp. 4, 310	pp. 292, 294	pp. 6, 309, 310	pp. 208, 296, 305	pp. 29, 309	p. 203
Provision for Self and Others	pp. 5, 10, 27	pp. 200, 202, 288, 232, 234, 254, 268	pp. 22, 26	pp. 236, 291	p. 38	pp. 183, 193, 248
Deception/Honesty	pp. 40, 49, 72, 134	pp. 154, 223, 244, 247, 261, 276, 281, 322, 358, 361, 369	pp. 23, 25, 79, 92, 135	pp. 296, 308, 321, 342	pp. 40, 72	pp. 206, 273, 293, 253, 277, 311
Survival Skills (sampling for Katniss—in arena—far too many to list all)	pp. 57, 89, 94, 96, 108	Gathering food: pp. 155, 170, 171, 196, 198 Combat and Hunting: pp. 151, 283, 174, 185–190, 250, 278 Medical Skills: pp. 178, 265 Observation: pp. 217, 228, 355	p. 96	pp. 260, 273, 313, 316, 319, 331, 338, 343	pp. 41, 89, 90, 95, 96, 108, 160	p. 250
True Emotions	p. 112	pp. 281, 243, 301	p. 133	pp. 316, 323, 343, 344, 369, 372	pp. 130, 135	p. 300

to note that all representations are accepted as long as they can be justified. No two teams or two individuals would have identical depictions. After all the hexagons have been arranged and explained with textual evidence from the novel, students place labeled arrows to denote the crosscutting science concepts (CCCs), again analyzing and justifying why that particular CCC is represented. Discussion of the rationale and evidence should not be rushed, as it encourages students to validate their reasoning. Figure 9.5 illustrates a sample hexagonal thinking activity created using Google Slides, although paper copies of the hexagons may also be used.

Each student then selects at least one CCC on which to write a justification using textual evidence. Additionally, students should be encouraged to capture their hexagonal projects with a screenshot, photograph, or by gluing paper on posters to present their thinking to other groups.

EXTENSION ACTIVITIES

Creating with Engineering Design (ED) Activities

Students can be challenged with any number of ED activities to apply and use content learned from *The Hunger Games*. The ED process includes a cycle of asking questions or defining a problem, brainstorming possible solutions, planning a model and making a materials list, creating a prototype, improving the final design as needed, and presenting the prototype to an audience. Table 9.3 describes an ED challenge for teams of students to design an invention for a

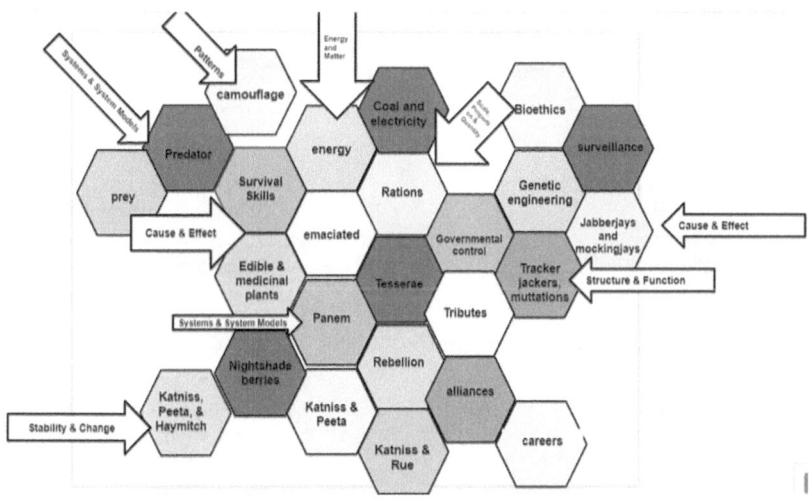

Figure 9.5. Hexagonal Thinking with *The Hunger Games* (created by authors).

Table 9.3. Engineering Design Process: Biomimicry Invention Activity

1. Ask or Identify the Problem	**Invention Supplies List**
The Capitol has asked your team to design an invention for your assigned district of Panem. Your invention needs to specifically help address your district's purpose (i.e., District 8 Textiles; District 10 Livestock, etc.). Your design needs to be inspired by nature and use biomimicry in terms of aesthetics, structure, or function. The Capitol will reward the teams with the most effective inventions incorporating biomimicry with additional food and supplies for the next year. You have a bag of supplies to work with and you must use these to create the prototype for your invention.	**(sample items)** Plastic jugs (milk cartons, etc.) Plastic straws and string Plastic plates and cups Plastic utensils Plastic holders Plastic packing material Plastic eggs Paper towels and toilet rolls Paper plates Newspaper Cardboard boxes
2. Imagine—Brainstorm ideas and choose one.	Steel cans
Conduct research into and take notes on examples of biomimicry. Come up with a list of ideas as a group that you would like to create for your invention.	Aluminum pie pans Soda cans Q-tips and cotton balls Pony Beads
3. Plan—Draw a diagram and gather needed materials.	Feathers Pipe cleaners = =
Before you construct, sketch your plan for the invention and gather additional supplies that you might need. You can use up to 1 meter of masking tape and/or string or any combination of the two. Develop a detailed drawing showing your product or system from at least two perspectives and include a written description of how the design works.	Yarn/ribbon stuff CDs
4. Create—Follow the plan. Test it Out!	
Construct your prototype and test how it works. Your prototype needs to weigh at least 10 ounces and cost less than $100 to produce.	
5. Improve—Discuss what you can do to make your invention work better. Repeat steps 1–5 to make changes as time permits.	
***Hunger Games*: Panem Districts**	**Advertisement**
District 1: Luxury District 2: Masonry District 3: Technology District 4: Fishing District 5: Power District 6: Transportation District 7: Lumber District 8: Textiles District 9: Grain District 10: Livestock District 11: Agriculture District 12: Mining	Create an advertisement for your invention. Include the following information as a minimum: Descriptive name for your invention. A snappy slogan or use of propaganda, techniques (bandwagon, testimonial, etc.) What need(s) can be addressed by your invention?

(continued)

Table 9.3. (Continued)

	Concluding Writing Activity Write an essay about examples of how engineers have looked to nature to help find solutions to societal challenges (cite sources). Describe how your group used nature to construct your invention. Use persuasive writing techniques to convince the Capitol to select your district's prototype for production.

specific Panem district inspired by biomimicry. Other ED challenges that can be developed include designing a water filtration system, building a solar oven, or building a launcher similar to a slingshot or bow and arrow (for use with marshmallows or cotton balls).

Choosing Creative Challenges

Choice board menus are an excellent way to allow for differentiation and promote student motivation because the students have the option to select projects that have the most meaning to them. Place three to four students into a team and ask them to select projects from the choice board shown in table 9.4. The expectations for this project are for the teams to select three projects to make a "tic-tac-toe" by completing a row up and down, across, or diagonally. The project options provide opportunities for students to further explore the science concepts and life themes within *The Hunger Games* novel. Most of the options allow for the creation of a digital product which could be shared through social media which is a popular choice for information-consumption and communication among adolescents. Science concepts highlighted within the projects include global warming, bioethics, alternative energy sources, innovations in medicine, the Clean Air Act, genetic engineering, making a scientific argument through a claim, evidence, and reasoning (CER) statement, and STEM careers. Life themes highlighted within the projects include exploring customs and lifestyles through the creation of a district token or a board game. The projects also explore social emotional learning through the options of formulating the team's thoughts about family, government, money, or competitions into the medium of a song or illustrating how a character was feeling during certain scenes within a cartoon. Completing this choice

Table 9.4. Choice Board Menu Extension for *The Hunger Games*

Original Song and Music Video	Screencast	District Token
Refer back to the prompts given in the Anticipation Guide for *The Hunger Games*. As a team of three to four select one on the prompts (e.g., Life is easier when you have money) to write your consensus statement. Put your statement in the form of a song—reword a popular children's song or other song if desired and video your team singing the song. Use a video editor of your choice to add transitions, headings, and music to your video creation.	Create a slide presentation about one of the following topics: Global Warming, Bioethics & CRISPR, Alternative Energy Sources, or Innovations in Medicine. Compare and contrast your selected topic to occurrences in *The Hunger Games* with historical as well as current events as applicable. Include at least ten slides with images and references. Record a five to ten-minute screencast of your presentation.	Katniss wore a mockingjay pin given by her friend Madge as her district token which served as inspiration and fond memories from home. Create a token that you would want to take with you to *The Hunger Games* if you were selected as a tribute. Write a one-paragraph description of what your token means to you. Design your token using Tinkercad and if you have access to a 3D printer, print your token.
Public Service Announcement	**Free Choice**	**Cartoon**
Read about the Clean Air Act and explore the pros and cons of fracking oil and mining coal. Write and record a Public Service Announcement for Mine Safety, Fracking Oil, or the Clean Air Act.	Design your own Activity! Need teacher approval first.	Select one character from *The Hunger Games* and illustrate a 5-frame cartoon summarizing main events. Synthesize what the character must have been thinking and feeling in your selected scenes.
Flipgrid and Claim Evidence Reasoning (CER) Statement	**Survivor Game**	**Children's Book**
Contribute to a Flipgrid post responding to one of the following claims, "Genetic Engineering is" . . . "good for society" or "detrimental for society." Provide evidence for your claim from at least three credible sources. Add reasoning for your claim specific to science concepts related to genetics. Attach a one-page typed summary of your CER to your Flipgrid post with references.	Create a board game that has the participants travel through Panem, including each district and the Capitol. Integrate survival skills, lifestyles, and customs that illustrate the way of life in each district (e.g., District 11 orchards & District 12 mines). Add hazards such as reaping, tracker jackers, and jabber jays.	Select one STEM career represented within *The Hunger Games* and write and illustrate a children's book for upper elementary students. Consider the following careers: engineer (focus on a particular type such as mechanical or civil), geneticist, geologist, pharmacist, ecologist, artist, or another of your choosing. Use Book Creator online to publish and share your book. Include text, images, audio, weblinks, and video in your book.

board focuses primarily on the SEPs of constructing explanations, engaging in argument from evidence, and obtaining, evaluating, and communicating information.

Extending Science Content beyond Life Science with *The Hunger Games*

Earth Science topics include observing the resources, soil types, and Earth processes such as weathering and erosion of the districts of Panem, and making comparisons to the geography and geology of North America. Studying the history and current state of the districts of Panem can be aligned with global warming and the environmental impact of using alternative energy sources (mining and fracking). Additionally, a number of characters in the book determined the time of day from the position of the sun and used the behavior of light and shadows as a means of survival whether it was hunting for food sources or for camouflage as a tribute within *The Hunger Games*.

Physical Science within *The Hunger Games* novel addresses motion and energy concepts with the use of weapons such as the bow and arrow, slingshots, and launchers for both hunting in District 12 for food and within the arena for food and/or other tributes. Chemistry was also a focus as mentioned previously with the use of medicinal herbs and creating potable water through filtration systems. Cinna, a costume designer, created synthetic fire for Katniss' and Peeta's costumes for the opening ceremonies leading to Katniss being known as the "girl on fire." The fire theme continued in the arena when the gamemakers used engineered fire to cause the tributes to move out of hiding.

Genetic Engineering Paired Texts

The Hunger Games novel brings to light the use of transgenic or genetically modified and engineered organisms as weapons within the world of Panem including the jabberjay, tracker jacker, and wolf-mutt. These genetically altered animals were called mutations in the text and are described in table 9.5.

These organisms are excellent for initiating discussion regarding the following questions: *what are the values and concerns of creating genetically modified organisms (GMOs) and how is bioethics considered or not considered in both* The Hunger Games *and in contemporary society?*

After reading the novel, these highlighted excerpts could be used as a starting point for studying genetics, genetic engineering, and GMOs with the use of additional paired texts and resources. Students can participate in using

Table 9.5. Genetically Modified Organisms in *The Hunger Games*

GMO	Description of GMO in Collins (2008)	Use	Representation
mockingjay	Crossbreed of a mutation, the jabberjay, created by the Capitol to spy on enemies and record their conversations and a mockingbird; districts used the jabberjay against the Capitol to spread misinformation. (pp. 42–43)	Reproduce both bird calls and human songs in melody.	Symbol of hope for Districts and Katniss.
tracker jacker	Capitol created mutation of giant gold wasps that will hunt down and kill anything that disrupts their nest; creates strong hallucinations that can cause madness or death. (pp. 185–186)	Protected strategic locations during the war; now wild.	Symbol of the Capitol's power over all living things even after the war.
mutation	Capitol designed mutation of Tributes that perished in the arena and are reincarnated as vicious, huge wolves. (p. 331)	Increased fear and tension for remaining Tributes in the arena; forced them to action.	Symbol of the corrupt power of the Capitol to manipulate and control citizens.

simulations to discover more about DNA and genetic engineering through becoming a virtual bioengineer (Amino Labs, 2022). Table 9.6 offers a list of informational text articles and books that share the pros and cons of genetic engineering that can be used as supplemental reading. Additionally, Marvel's X-Men comics and movies feature mutants who were either born with naturally occurring genetic mutations that gave them superpowers or, as in the case of Wolverine, were produced with experimental genetic engineering.

After reading and exploring different plant and animal GMOs, students should be prepared to create annotated drawings to describe what happens genetically when GMOs are created such as plants, insects, or humans. Students can create bioethics position statements pertaining to the current status of genetic engineering. They should consider how responsible Panem was in genetically engineering the jabberjays, tracker jackers, and wolf-mutts as well as how responsible real-life scientists have been in engineering organisms.

Table 9.6. Genetic Engineering: Informational Texts and Books

Source	Summary	Major Concepts
McClatchy Foreign Staff. (2013, October 17). Suicide mosquitoes a gene-altered weapon in war against dengue fever. *Newsela*.	Public health officials release gene-altered mosquitoes that produce non-viable offspring in an effort to prevent mosquito-borne illnesses such as dengue fever.	GMOs, public health, minimal ecosystem impact
Associated Press. (2020, March 23). Doctors try first CRISPR editing in the body for blindness. *Newsela*.	A patient was recently treated for an inherited form of blindness using the first known DNA editing procedure with CRISPR inside the body.	GMOs, public health, governmental oversight
Issacson, W. (2021). *The code breaker: Jennifer Doudna, gene editing, and the future of the human race.* Simon & Schuster.	A popular biographer presents the life and work of Doudna from being told that girls don't do science to the gritty work in the lab and the moral consequences of gene alteration.	GMOs, governmental control, ethics
Pahara, J. & Legault, J. (2021). *Zero to genetic engineering hero: The beginner's guide to programming bacteria at home, school, & in the makerspace.* Make: Community, LLC.	This hands-on guide is designed to provide both content and hands-on activities for learners to explore genetic engineering and a "Biology-as-a-Technology" mindset.	GMOs, technology, ethics, hands-on activities
Ridge, Y., & Boersma, A. (2020). *CRISPR: A powerful way to change DNA.* Annick Press.	Written for secondary school students, this text presents science with detailed illustrations and thought-provoking questions about the ethics of using CRISPR.	GMOs, ethics

CONCLUSION

The Hunger Games has been an extremely popular dystopian novel for young adults and has excellent potential as a motivational hook for students in secondary science classrooms. After reading the novel for pleasure, students can initially make connections with the life themes presented including developing and strengthening relationships with family and friends, similarities and differences between cultures and customs within our world, the need to belong with a group, and competition and the importance of teamwork. Depending upon the socioeconomic status of the students, they will have varying connections to the survival skills needed, particularly hunting and gathering food and access to basic needs for survival: food, water, and shelter.

Science teachers can help students make explicit connections between concepts in the novel and the content within the science classroom. Of particular importance is providing an authentic context in which students explore how science works, specifically using research in order to make informed decisions and construct explanations or participate in meaningful scientific argumentation. The science concepts incorporating biotic or natural and abiotic or geopolitical resources, genetic engineering, bioethics, alternative energy sources, innovations in medicine, and survival needs among others are represented within *The Hunger Games* which provides a meaningful context and starting point to make connections with our real world.

REFERENCES

Amino Labs. (2022). *Virtual bioengineer simulations.* https://amino.bio/pages/vbioengineer.

Collins, S. (2008). *The hunger games.* Scholastic Press.

Johnson, D. D., & Pearson, P. D. (1978). *Teaching reading vocabulary.* Rinehart and Winston.

National Research Council. (2012). *A framework for K-12 science education: Practices, crosscutting concepts, and core ideas.* The National Academies Press.

Photos for Class. (2022). https://www.photosforclass.com/.

Rollins, S. P. (2014). *Learning in the fast lane: 8 ways to put all students on the road to academic success.* Association for Curriculum and Supervision.

Yell, M. M., & Scheurman, G. (2004). The anticipation guide: Motivating students to find out about history. *Social Education, 68*(5), 361–363.

Index

5E learning cycle, ix, 24, 25, 28, 39, 41, 42

adaptation(s), x, 83, 118–122, 124, 128–131
alien plant(s), 46, 49–51, 53, 54
alternative energy sources, 169–71, 174
apocalyptic, 46, 59

bioethics, 134, 158, 159, 169–72, 174
biology, ix, x, 71, 85, 88, 116, 133, 135, 147, 154, 173; conservation, 61; dehydration, 3, 11; human, 1; plant, 45–48, 51, 54, 57
biomimicry, 168, 169
botany, 61–65, 68, 75, 76
Bryan, William Jennings, 116, 123, 124, 125

carnivorous plants, 51
cells, 49, 69–71, 103–5, 138–140, 142
cell theory, 69–71, 76
Centers for Disease Control and Prevention, 95
classification, 64, 68, 76
climate, 5, 27–30; change, ix, 1, 3, 6, 23–43, 61, 83, 149, 153; crisis, ix, 20, 23–26, 27–30; political, 111
COVID-19, x, 95–98, 103, 107, 111, 112

cryptogenic, 49, 51, 52, 56, 57

Darrow, Clarence, 116, 123–125
Darwin, Charles, 82, 87, 118–120, 126, 131
Darwin's theory of evolution, x, 78
data, 23, 27, 29, 30, 32, 34, 39, 40, 52, 73, 95, 97, 117, 135; analysis, ix, 24, 28, 34, 35, 37, 38, 40, 42, 79, 83, 88, 131; base, 126; collection, 24, 28, 119; crowd-sourced, 59; interpreting, 73, 79, 83, 88, 152; literacy, 24, 33, 41; patterns, 34; synthesis, 54
dehydration, 2, 3, 6–8, 11, 12, 14, 15, 98
disease(s), x, 95–102, 104, 107–10, 112, 157
DNA, 49, 70, 88, 133, 136, 138–143, 154, 157, 158, 172, 173
drought, ix, 1–6, 18, 20, 23, 24, 25, 30–32, 42, 145, 147

ecology, ix, 1, 8, 64, 73, 158, 159; fire, ix, 1
environmental literacy, ix, 24, 42
environmental science, 29, 32, 61, 88, 153
ethics, ix, x, 10, 133, 134, 135, 140, 141, 148, 150, 152, 153, 157, 173

etymology, 67, 68
evolution, x, 78, 115–124, 127, 128, 129, 131, 132, 151, 152; biological, 61, 89, 90
explainer creation(s), 92

first lines, 81–84
food shortages, 141, 151

genetic(s), ix, x, xi, 64, 133, 134, 136, 138, 145, 147, 149, 151, 152, 154, 158, 159, 170, 171; code 142–145, 147; engineering, 133, 134, 151, 154, 157, 158, 169–74; material, 49, 140; modification/manipulation, 140, 141, 144, 145, 146, 147, 152, 154, 156, 157; mutation, 49; research, 153, 157
global warming, 23, 25, 28, 35, 42, 169–71

human impact, ix, 23–25, 30, 32, 41, 61

immune system, x, 96, 99, 101, 104–7, 109
influenza, 97, 101, 103; Spanish Influenza, x, 96, 97, 111, 112; vaccine(s), 107
infographic(s), 39–41
invasive plant species, ix, 46, 49, 50, 54, 58, 59

natural disasters, 1, 3, 6, 7, 10, 16, 17, 27

naturalist notebooks, 85, 86, 89, 92
natural selection, 83, 89, 118, 120, 126, 128

pandemic(s), x, 95–98, 101, 103, 110–12
patterns, ix, 1, 20
plant(s): awareness disparity, 65, 76; biology, 45–48, 51, 54, 57; blindness, 45, 65
poison/poisonous, 65–68, 72, 73

scientific observation, 48, 80, 81, 85, 86, 89, 92
Scopes Trial, 116, 121–125, 127, 128, 129, 131
sea level rise, 24, 25, 30, 32, 35–40, 42
STEM, 24, 72, 74, 76, 92
survival skills, xi, 19, 158, 166, 170, 173

Tate, Calpurnia, x, 77, 78, 80–85, 88–90, 93

water: collection, 8, 19; conservation, ix, 1–4, 6, 7, 14, 16–20; filtration, xi, 158, 169, 171; sources, 3–6; storage, 8; systems, ix, 1–4, 7, 8, 10, 17–20; treatment, ix, 1–4; zombies, 2, 7, 11, 14
weather, 5, 24–33, 40, 42, 85, 92; cycle, 23
wildfires, 1, 3, 5, 7, 15–17, 20, 23, 25, 42

About the Contributors

Julie C. Baker, Ph.D., is a full professor and serves as the associate dean in the College of Education at Tennessee Tech University in Cookeville, Tennessee. She holds degrees in Secondary Education (B.S.), Instructional Leadership (M.A.), and a Ph.D. in Exceptional Learning with a concentration in Literacy, all from Tennessee Tech. In addition to her role as a classroom teacher and literacy coach at two large Tennessee high schools, she has taught a variety of university-level courses focused on both qualitative and quantitative research methods, rural schools and communities, secondary education, service learning, and more.

Kathryn Baldwin, Ed.D., is an associate professor of science education at Eastern Washington University. Dr. Baldwin teaches coursework in science methods and environmental and sustainability education. Her research interests focus on science education, earth, and environmental education, outdoor learning, science teaching self-efficacy, and problem and project-based learning.

Ashley S. Boyd, Ph.D., is a former secondary English teacher and associate professor of English education at Washington State University. She teaches English methods, young adult literature, and critical theory, and her current research focuses on secondary teachers' social justice pedagogies and students' development of social action projects.

Chris Cook, Ph.D., is a professor of middle-level education in the Department of Curriculum & Instruction at Appalachian State University. He is a former middle school science teacher. His research focuses on middle-level teacher preparation and middle-level pedagogy. He received his Ph.D. in curriculum and instruction from the University of North Carolina Greensboro.

Katharine Covino, Ed.D., is an associate professor of English studies and teaches writing, literature, and teacher-preparation classes at Fitchburg State University. Three areas of current scholarship focus on (a) critical pedagogy in literacy-learning classrooms, (b) applying indigenous lenses to examine cultural myths, and (c) collaborating with middle, secondary, and postsecondary teachers on issues related to literacy praxis. Prior to teaching at the university level, she taught middle school and high school in Austin, Texas. She is also a children's book author. Two recent books seek to support children and their families in understanding, processing, and addressing the challenges of living through a global pandemic in a funny, accessible, kid-friendly way. An upcoming book uses humor and science to help kids use and not abuse digital technology.

Janine J. Darragh, Ph.D., is a former high school teacher of thirteen years and current associate professor of literacy and ESL at the University of Idaho where she instructs courses on adolescent literature, English teacher preparation, and ESL. Her current research focus is sociocultural issues in English teaching and learning in both the United States and abroad and supporting teachers and learners in contexts of crisis.

Michael DiCicco, Ph.D., is an associate professor of Literacy Education in the Department of Teacher Preparation and Educational Studies at Northern Kentucky University. He is a former middle school language arts teacher. His research focuses on middle-level teacher preparation and middle-level literacy. He received his Ph.D. in curriculum and instruction from the University of South Florida.

Frances A. Hamilton, Ed.D., is an assistant professor of elementary education at the University of Alabama in Huntsville. Her current research interests include integrating science with language arts.

Heather Johnson, Ph.D., is an associate professor of science education in Peabody College's Department of Teaching and Learning at Vanderbilt University, Nashville, Tennessee. Her scholarship centers around supporting secondary science teacher candidates and Inservice teachers in learning how to surface, recognize, and leverage students' resources and lived experiences in the world as assets for science learning in classrooms. By positioning student thinking and student resources at the center of their practice, teachers will move their practices to be more ambitious, equitable, and inclusive. Framing her practice and research through this lens, she often uses video analysis as a pedagogy to explore what ambitious, equitable, and inclusive science teaching and learning environments and practices look like and reflect with teachers on how to shift current practices toward this vision.

Shawn E. Krosnick, Ph.D., is an associate professor of biology and curator of the Hollister Herbarium in the College of Arts and Sciences at Tennessee Tech University in Cookeville, Tennessee. She earned her B.S. in plant science from Cornell University and a Ph.D. in plant systematics from the Ohio State University. She teaches courses including General Botany, Plant-Animal Interactions, and Plant Anatomy. She also advises graduate and undergraduate students conducting research on plant reproductive biology, systematics, and taxonomy. She cares deeply about making botany fun and engaging for her students and addressing diversity, equity, and inclusion in all aspects of her job.

Ben Lawhorn is a ninth-grade biology teacher at Winchester High School in Winchester, Massachusetts. He holds an undergraduate degree in integrative biology and a master's degree in education with a specification in High School Biology and has experience in researching marine life. He specializes in engaging students with outdoor learning experiences.

Kelly Moore, Ph.D., is a senior lecturer in the Department of Curriculum and Instruction, College of Education at Tennessee Tech University in Cookeville, Tennessee. She has been working in science education for over twenty years, both in the high school setting and in higher education. Her current responsibilities include working with preservice science teachers through their methods courses and field experiences. Kelly is passionate about helping science teachers create engaging, student-centered classrooms that utilize authentic scientific practices.

David Nurenberg, Ph.D., is an associate professor of education at Lesley University in Cambridge, Massachusetts, who has taught high school humanities for over twenty years. He is the author of *What does injustice have to do with me? Engaging privileged white students with social justice* (Rowman & Littlefield, 2020), hosts the progressive education podcast Ed Infinitum (www.ed-infinitum.com), and consults with schools around student-centered and project-based learning methods.

Amy Palmeri, Ph.D., is an associate professor of elementary education in the Department of Teaching and Learning at Vanderbilt University, Nashville, Tennessee. Her teaching and research interests focus on the development of teacher candidates within the context of initial teacher preparation. A particular interest focuses on teacher candidates' developmental trajectories related to disciplinary understandings and practices within science and social studies alongside the developmental trajectories of related pedagogical knowledge and skill. Of central interest is coming to understand how these two distinct developmental trajectories interact and intersect as teacher candidates' grow and develop across their preparation program. Further, she explores role that teacher educators play in mediating teacher candidate learning.

Emily Pendergrass, Ph.D., is an associate professor of literacy education. Education in Peabody College's Department of Teaching and Learning at Vanderbilt University, Nashville, Tennessee. She teaches literacy development courses with social justice, critical lens. Her research interests include improved reading practices with adolescents who struggle with school-based reading and how these practices intersect with teacher decision-making. Additionally, Emily works closely with the local, high-priority public schools, other schools around the United States, and non-profit agencies in literacy coaching and facilitating professional development workshops. Her goal is to enhance literacy instruction for all learners. In this role, she works with inservice and preservice teachers to think about literacy as a social justice initiative where all students have a civil right to access reading, writing, and critical thinking, and to engage in high-quality learning that will enhance their lives.

Kristen Pennycuff Trent, Ph.D., is a professor of literacy education at Tennessee Technological University. After teaching elementary school for over six years, she has spent the last twenty years working with undergraduate and graduate programs both on the main campus and in the 2+2 program. As a grant writer, Pennycuff Trent has been awarded over 2.4 million dollars for her work with literacy professional development for Pre-K-12 educators. Funded projects have ranged from an early childhood education program for at-risk families from the TN Department of Education, Reading Excellence Act and Reading First grants, Tennessee Higher Education Commission Improving Teacher Quality grants, and Math and Science Partnerships projects in science literacy.

Kristina L. Podelnyk is an NBPTS-certified teacher in secondary science with over a decade of experience working in middle, junior, and high school classrooms. Certificated in biology, chemistry, Project-Based Learning, and STEM education, Kristy loves designing and implementing classroom experiences that feature diverse voices, new perspectives, and seize on opportunities for authentic student-directed engagement.

Erin Rehrig, Ph.D., holds a doctoral degree in plant science from the University of Missouri, an M.Ed. in science education, an M.S. in horticulture from Penn State and a B.S. in biology from the Bloomsburg University of Pennsylvania. Her research interests include plant-insect interactions, plant responses to biotic and abiotic stresses, and the use of C.U.R.E.s (course-based undergraduate research experiences) in plant biology laboratories. She teaches introductory biology, biochemistry, science education, and plant biology classes and is currently serving as interim chair of the Biology and Chemistry Department at Fitchburg State University. She lives in Massachusetts with her husband, two children, and a very spoiled Labrador mix.

Shelly Shaffer, Ph.D., is an associate professor of literacy in the School of Education at Eastern Washington University, where she has worked since 2015. Dr. Shaffer's primary research focus is the study and teaching of Young Adult literature. She has published several book chapters focused on teaching and analyzing Young Adult literature and coedited the book *Contending with Gun Violence in the English Language Classroom*, focused on teaching strategies for addressing (and talking about) gun violence. Her current scholarly interests are teaching Young Adult Literature, research in the teaching of Young Adult literature, critical literacy, Young Adult literature in the content area, and reading motivation.

Dana L. Skelley, Ed.D, is an assistant professor of literacy education at the University of Alabama in Huntsville. Her current research interests include afterschool programs and digital literacies.

Leslie Suters, Ph.D., is a professor in the College of Education's Curriculum and Instruction Department at Tennessee Technological University. She teaches elementary mathematics, science, and technology methods courses to K-5 preservice teachers in TTU's 2+2 program. Her research interests include collaborative professional learning opportunities for teachers with a focus on computational thinking within STEM courses. She has developed, led, and participated in numerous multi-year STEM professional development programs for K12 educators with external funding through federal (Race to the Top) and state (Math Science Partnership, Improving Teacher Quality) programs.

About the Editors

Paula Greathouse, Ph.D., has coedited several books including *Adolescent Literature as a Complement to the Content Areas* (Rowman & Littlefield) book series, *Queer Adolescent Literature as a Complement to the English Language Arts Curriculum* editions 1 and 2 (Rowman & Littlefield), *Breaking the Taboo with Young Adult Literature* (Rowman & Littlefield), *Young Adult and Canonical Literature: Pairing and Teaching* (Rowman & Littlefield, 2021), and *Shakespeare and Young Adult Literature: Pairing and Teaching* (Rowman & Littlefield, 2021). Her research on adolescent literacy and young adult literature has been published in books and top-tier journals such as *Educational Action Research, Study and Scrutiny: Research on Young Adult Literature, The Clearing House*, and *English Journal*. She was a secondary ELA and Reading educator for sixteen years. She has won the Florida Council of Teachers of English High School Teacher of the Year and the National Council of Teachers of English (NCTE) Teacher of Excellence awards.

Melanie Hundley, Ph.D., is a professor in the Practice of English Education at Vanderbilt University's Peabody College. She teaches writing methods courses that focus on digital and multimodal composition and young adult literature courses that explore race, class, gender, and sexual identity in young adult texts. Her research and work on young adult literature have been published in *Young Adult and Canonical Literature: Pairing and Teaching, Shakespeare and Adolescent Literature: Pairing and Teaching, Participatory Literacy in P-12 Classrooms in the Digital Age, Breaking the Taboo with Young Adult Literature, Teaching Young Adult Literature Today: Insights, Considerations, and Perspectives for the Classroom Teacher, Using Children's Literature*, and *Perspectives on Digital Comics*. She has also published in *The ALAN Review, The English Record, Language Arts*, and *Journal of Adolescent and Adult Literacy*. She is past editor of *The ALAN Review*. She is currently the Coordinator of the Secondary Education English Education program and Associate Chair for the Department of Teaching and Learning at Vanderbilt University's Peabody College.

Stephanie Wendt, Ed.D., is an associate professor of teacher education at Tennessee Tech University. She teaches science methods, educational data assessment, field experience, and graduate courses in learning theory and grant writing. She is a member of the Tennessee Science Education Leadership Association and serves on the Tennessee Science Teachers Association Board of Directors. She leads professional development for K-12 science educators at the state, national, and international levels. Her recent authorship includes publications in *The International Journal of Science, Mathematics and Technology Learning* (2022) and the KDP's *The Teacher Advocate* (Summer, 2021). Dr. Wendt has book chapters published in *Adolescent Literature as a Compliment to the Content Areas for Science and Math* (Rowman & Littlefield, 2017), *Bringing STEM to the Elementary Classroom* (2016), and *Integrating 3D Printing into Teaching & Learning: Practitioners' Perspectives* (2020). She is also a Senior Reviewer for the IAFOR *Journal of Education: Studies in Education* and NSTA's award-winning journal *Science and Children*.

www.ingramcontent.com/pod-product-compliance
Lightning Source LLC
Chambersburg PA
CBHW020125240426
43673CB00038B/591